BBNJ
国际协定谈判中国代表团发言汇编

（一）

郑苗壮　等◎编

中国社会科学出版社

图书在版编目(CIP)数据

BBNJ国际协定谈判中国代表团发言汇编．一／郑苗壮等编．—北京：中国社会科学出版社，2019.8

ISBN 978-7-5203-4847-8

Ⅰ.①B…　Ⅱ.①郑…　Ⅲ.①国际海域-海洋生物资源-资源保护-国际条约-文件-汇编　Ⅳ.①P745②D993.5

中国版本图书馆CIP数据核字(2019)第171401号

出 版 人　赵剑英
责任编辑　梁剑琴
责任校对　闫　萃
责任印制　郝美娜

出　　版　中国社会科学出版社
社　　址　北京鼓楼西大街甲158号
邮　　编　100720
网　　址　http：//www.csspw.cn
发 行 部　010-84083685
门 市 部　010-84029450
经　　销　新华书店及其他书店

印刷装订　北京市十月印刷有限公司
版　　次　2019年8月第1版
印　　次　2019年8月第1次印刷

开　　本　880×1230　1/32
印　　张　8.25
插　　页　2
字　　数　215千字
定　　价　58.00元

凡购买中国社会科学出版社图书，如有质量问题请与本社营销中心联系调换
电话：010-84083683

编委会（按姓氏笔画排序）：

马新民　张海文　陈　越　贾　宇

主　编：郑苗壮　刘　岩　裘婉飞　唐冬梅

参加人（按姓氏笔画排序）：

王鹏超　卢晓强　冯　军　朱　璇
刘　洋　汤熙祥　李琳琳　邱雨桐
张小勇　张继伟　张　爽　陈建明
陈建忠　罗　阳　赵　畅　徐金钟
徐敏婕　高　岩　唐　议　董　跃
蒋小翼　景　辰　阙占文　管毓堂
谭　伟

序

国家管辖范围以外区域海洋生物多样性（BBNJ）养护和可持续利用问题国际文书谈判政府间大会于2018年9月正式开启，标志着有关BBNJ的国际立法进程进入新阶段。各国正朝着在《联合国海洋法公约》框架下制定第三个执行协定的方向努力迈进。

BBNJ国际文书谈判是当前海洋法领域最重要的国际立法进程，涉及海洋遗传资源的获取和惠益分享、海洋保护区等划区管理工具、环境影响评价、能力建设和海洋技术转让等重要问题，攸关海洋新资源和海洋活动空间等战略利益，牵动国际海洋秩序的调整和变革，各国对国际文书谈判予以高度重视。

学界和实务界关于BBNJ问题的讨论由来已久，2004年该问题被正式纳入联合国议程，同年第59届联大设立BBNJ问题特设工作组，讨论和研究加强BBNJ养护和可持续利用的可行方案。历经11年九次特设工作组会议的讨论磋商，联大于2015年6月19日通过69/292号决议，决定在《联合国海洋法公约》框架下就BBNJ问题制订具有法律约束力的国际文书，并建立谈判预备委员会，就国际文书草案的要素进行讨论并向联大提出实质建议。预备委员会历时两年，先后举行四次会议，于2017年7月通过了向联大提交的要素建议草案，同时声明该草案不代表各方就相关要素达成的共识，其不影响各国在未来政府间大会谈判中的立场。在此基础上，联大于2017年12月24日通过第72/249号决议，决定在联合国主持下于2018年至2020年上半年首先召开四次政府间大会，就制订有关国际文书展开政府间谈判。

中国作为快速发展中的海洋大国，高度重视国家管辖范围以外区域的海洋生物多样性国际立法问题，积极参与了历次特设工作组会议、谈判预备委员会会议以及政府间大会首次会议。我本人作为中国代表团团长以及预备委员会和政府间大会首次会议的主席团成员，参加了预委会所有四次会议和政府间大会首次会议。会上，中国代表团深度参与了 BBNJ 有关问题的讨论和磋商，积极发出中国声音，努力贡献中国智慧和方案，在不少问题上发挥引领作用，特别是在 2018 年 9 月政府间大会首次会议上的发言全面而富有建设性，有效扩大了中国在有关进程中的话语权和影响力。中国代表团还两度以中国政府名义就国际文书草案要素提交书面意见，系统阐述中国对有关问题的立场主张和法律依据，对中国参与和引导会议进程发挥了重要作用。

在前一阶段参与相关磋商和谈判过程中，包括自然资源部海洋发展战略研究所同仁在内的一批国内专家学者组成强有力的研究团队，对 BBNJ 国际文书所涉法律问题开展深入研究，积极建言献策，为中国政府代表团参与有关谈判提供了有力支撑。为了帮助学界和实务界及时跟踪了解相关立法进程，海洋发展战略研究所的同事们对中国代表团在前一阶段历次会议上的发言以及中国政府书面意见等文件进行了汇编，旨在反映中国参与 BBNJ 国际文书谈判进程的阶段性情况，以推动相关问题的深入研究。

相信本书对于国内学界和实务界进一步研究 BBNJ 问题以及中国政府代表团参与后续政府间谈判具有重要的参考价值。在此，谨对有关同事的努力和付出表示衷心感谢！

马新民①

2019 年 1 月 17 日于北京

---

① 作者系时任外交部条法司副司长、BBNJ 预委会和政府间大会第一次会议中国代表团团长，现任中国驻苏丹大使。

目　录

中国政府关于国家管辖范围以外区域海洋生物多样性养护和可持续利用问题国际文书草案要素的书面意见

中华人民共和国政府（中国政府）高度重视国家管辖范围以外区域海洋生物多样性（BBNJ）的养护与可持续利用问题，派代表团积极参与了历次 BBNJ 特设全体非正式工作组会议和预备委员会（预委会）前三次会议。中国政府愿在后续磋商中继续发挥建设性作用，为国际社会更好养护和可持续利用 BBNJ 贡献力量。

中国政府支持 77 国集团加中国在前三次预委会期间发表的有关意见以及就 BBNJ 问题提交的书面意见，并愿在此基础上以政府名义表达如下看法。本书面意见不妨碍中国政府在今后讨论中进一步提出意见和建议。

一、基本立场

中国政府支持联合国大会（联大）通过题为“根据《联合国海洋法公约》的规定就国家管辖范围以外区域海洋生物多样性的养护和可持续利用问题拟订一份具有法律约束力的国际文书”的第 69/292 号决议，强调预委会的职权是“就根据《公约》的规定拟订一份具有法律约束力的国际文书的案文草案要点向大会提出实质性建议”。有关各方应严格按照决议授权开展相关工作，预委会最终所提实质性建议应尽最大努力在协商一致的基础上反映各方共识。同时，中国政府支持预委会以 2011 年各方达成的共识为基础，“一揽子”同步处理海洋遗传资源包括惠益分享、划区管理工具包括海洋保护区、环境影响评价、能力建设和海洋技术转让等问题。

第一，新国际文书是在《联合国海洋法公约》(《公约》) 框架下制定的国际法律文件，应符合《公约》的目的和宗旨，应是对《公约》的补充和完善，不能偏离《公约》的原则和精神，不能损害《公约》建立的制度框架，不能损害《公约》的完整性和平衡性。各国根据《公约》在航行、科研、捕鱼等方面享有的自由和权利不应受到减损。《公约》关于沿海国的权利和义务的规定，包括对 200 海里以外大陆架的权利和义务的规定，不应受到影响。

第二，新国际文书不能与现行国际法以及现有的全球、区域和部门的海洋机制相抵触，不能损害现有相关法律文书或框架以及相关全球、区域和部门机构，特别是不能干预联合国粮农组织、区域渔业管理组织和安排、国际海事组织、国际海底管理局等机构的职权。新国际文书应促进与现有相关国际机构的协调与合作，避免职权重复或冲突。

第三，新国际文书的有关制度安排应有充分的法律依据和坚实的科学基础，并在 BBNJ 养护与可持续利用之间保持合理平衡。

第四，新国际文书应兼顾各方利益和关切，立足于国际社会整体和绝大多数国家的利益和需求，特别是应顾及广大发展中国家的利益，不能给各国尤其是发展中国家增加超出其承担能力的义务和责任。

二、海洋遗传资源包括惠益分享

国家管辖范围以外区域海洋遗传资源对于人类具有巨大实际或潜在价值。新国际文书的有关制度安排应有利于促进科研和鼓励创新，公平公正地分享养护和可持续利用国家管辖范围以外区域海洋遗传资源所产生的惠益，提升人类共同福祉。

（一）定义

中国政府认为《生物多样性公约》和《粮食和农业植物遗传资源国际条约》中关于遗传资源的定义是讨论新国际文书中海洋遗传资源定义的参考。新国际文书中关于海洋遗传资源的定义应包括以下四个要素：一是来自海洋的动物、植物和微生物或其他来源；二是含有遗传功能单元的遗传材料；三是具有实际或潜在价值；四是来源于国家管辖范围以外区域。

中国政府注意到，衍生物是生物化学合成的产物，不含有遗传功能单元，而且《生物多样性公约》和《粮食和农业植物遗传资源国际条约》关于遗传资源的定义本身都没有包括衍生物，新国际文书有关遗传资源的定义也不应包括衍生物。

中国政府认为，应对国家管辖范围以外海域的鱼类进行区分。对于作为商品的鱼类，《联合国海洋法公约》和《1982 年 12 月 10 日〈联合国海洋法公约〉有关养护和管理跨界鱼类和高度洄游鱼类种群的规定执行协定》（1995 年《鱼类种群协定》）等国际条约已经作出详尽规定，现有的区域渔业管理组织和安排基本覆盖了所有公海海域，养护和可持续利用渔业资源应该继续由现有的渔业组织或安排来管理，其不应作为新国际文书规范的事项。对于作为海洋遗传资源载体的鱼类，应遵循《公约》有关海洋科研自由的相关规定。

（二）获取

在预委会讨论中，有些国家提出海洋遗传资源获取包括原生境获取、非原生境获取和生物信息数据获取三种类型。原生境获取是指从国家管辖范围以外区域的自然环境中获取或采集海洋遗传资源。非原生境获取和生物信息数据获取，是对由原生境获取的海洋遗传资源进行实验室分离、鉴定、筛选、培养和计算机模拟分析后所得的资源、信息、材料和数据等的获取。中国政府认为，原生境获取活动本质上属于《公约》规定的国家管辖范围外区域的海洋科学研究，应适用自由获取制度，以促进海洋遗传资源的开发和可持续利用。

（三）惠益分享

国家管辖范围以外区域海洋遗传资源的采样、研发和商业化具有技术要求高、时间消耗长、资金投入大、结果不确定等特点。有关海洋遗传资源惠益分享机制应总体有利于 BBNJ 的养护和可持续利用，鼓励海洋科学研究，促进全人类对海洋遗传资源的惠益分享。中国政府认为，在充分照顾发展中国家关切和需求的前提下，预委会应优先考虑样本的便利获取、信息交流、技术转让和能力建设等非货币化惠益分享机制。同时，中国政府对探讨建立货币化惠益分享机制持开放态度。

三、划区管理工具包括海洋保护区

中国政府支持促进 BBNJ 的养护和可持续利用，高度重视划区管理工具包括海洋保护区。

（一）定义

划区管理工具包括多种管理形式和方法，不限于海洋保护区。中国政府认为，新国际文书所规范的划区管理工具定义应包括但不限于以下三个基本要素：一是目标要素，划区管理工具应旨在养护和可持续利用海洋生物多样性。二是地理范围要素，划区管理工具应仅适用于公海和国际海底区域。三是功能要素，划区管理工具应涵盖不同功能和管理方法。

关于海洋保护区的定义，中国政府认为，现行的《生物多样性公约》和一些区域性文书中的有关定义可以作为新国际文书对海洋保护区定义的参考。

（二）设立海洋保护区的原则和制度

《公约》序言明确规定："意识到各海洋区域的种种问题都是彼此密切相关的，有必要作为一个整体来加以考虑"。按照此种精神，在设立海洋保护区的原则和制度上，可考虑采取一体化海洋管理方法，以弥补目前分区域、按部门管理方法的不足。

对于新国际文书有关设立海洋保护区的具体指导原则，中国政府总体支持77国集团加中国在预委会此前会议上关于此议题的发言，并强调以下原则：一是必要性原则。海洋保护区是工具，而不是目标，建立海洋保护区应以确有必要为前提。二是比例原则。保护措施须与保护目标和效果相适应，在符合成本效益的前提下予以适用。三是科学证据原则。设立海洋保护区需有坚实的科学证据，评估受保护生态系统、栖息地和种群等的潜在威胁和风险。四是区别保护原则。按照不同海域、生态系统、栖息地和种群等各自的特点，适用不同的管理工具予以保护。五是国际合作原则。各国及国际组织应在建立海洋保护区问题上加强合作。

中国政府认为，设立海洋保护区需要满足一定的实体要件和程序要件。在实体要件方面，海洋保护区的设立应有明确的保护目标、确定的保护对象、具体的保护范围、适当的保护措施、合理的保护期限等。在程序要件方面，设立海洋保护区应遵循特定的程序，包括申请、咨询、审查、决策、管理和监督等。

根据联大第 69/292 号决议，新国际文书作为《公约》的执行协定，应旨在查漏补缺、填补空白。在海洋保护区等划区管理工具方面，无论新国际文书作出何种制度安排，其都不能影响国际海底管理局、国际海事组织、区域渔业管理组织等现有机构的相关职能以及现有相关国际条约的有关规定。但在设立海洋保护区的具体过程中，新国际文书有必要与现有区域或部门机构开展合作与协调。

新国际文书有关海洋保护区等划区管理工具的规定与现有相关制度的关系，涉及新国际文书与现有国际条约的关系问题，应遵循《维也纳条约法公约》有关条约适用的一般准则，予以妥善处理。

四、环境影响评价

中国政府高度重视海洋环境影响评价，认为新国际文书有关环境影响评价的制度安排应遵循《公约》所确定的基本法律框架和程序要素，同时顾及其他国际文书有关环境影响评价的规定。

《公约》第 206 条规定："各国如有合理根据认为在其管辖或控制下的计划中的活动可能对海洋环境造成重大污染或重大和有害的变化，应在实际可行范围内就这种活动对海洋环境的可能影响作出评价。"根据该条，中国政府认为，新国际文书有关环境影响评价的主体应是拟开展海洋活动的国家；对象应是各国管辖或控制下的计划中的"活动"，不包括战略环境影响评价；启动门槛应是"有合理依据认为""可能造成重大污染或重大和有害的变化”。另外，考虑到科学信息、技术方法、成本、能力等因素，累积影响评价是否“在实际可行范围内”值得商榷。

《公约》第 194 条第 2 款规定：“各国应确保……在其管辖或控制范围内的事件或活动所造成的污染不致扩大到其按照本公约行使主权权利的区域之外。”中国政府认为，《公约》对国家管辖范围内的活动的跨界影响已作出规定，新国际文书所规定的环境影响评价的范围应限于国家管辖范围以外区域的活动，包括可能对沿海国管辖海域产生重大环境影响的活动，而不应包括发生在国家管辖海域内的活动。

根据联大第 69/292 号决议，新国际文书作为《公约》的执行协定，不应损害现有国际法律文书或框架，以及全球、区域、部门机构。新国际文书有关环境影响评价的规定不应损害现有国际机构的职权以及现有国际文书的有关规定。中国政府认为，在公海深海捕鱼、国际海底区域深海矿产勘探、倾倒废物等领域已有的环境影响评价方面的规定不应受到损害。

中国政府注意到，有关国家提出制定环评对象活动类型清单。中国政府认为，各类海上活动对海洋环境的影响不尽相同，采取清单列举的方式有一定的局限性。如各方认为确有必要制定清单，中国政府认为，该清单应是开放性的、建议性的，不具有法律拘束力。

五、能力建设和海洋技术转让

能力建设和海洋技术转让是提升发展中国家养护和可持续利用 BBNJ 能力的重要手段，对实现海洋环境保护和可持续发展整体目标不可或缺。中国政府支持 77 国集团加中国关于能力建设和海洋技术转让的总体立场，愿补充如下意见：

第一，新国际文书关于能力建设和海洋技术转让方面的规定应以《公约》第十四部分有关规定为基础，并遵循针对性、有效性、平等互利、合作共赢等原则。

第二，新国际文书应充分照顾发展中国家的需要和利益，特别是小岛屿发展中国家、最不发达国家、内陆国和地理不利国以及有特殊利益需求的国家。

第三，新国际文书应鼓励通过多种形式的国际合作，加强发展中国家能力建设和海洋技术转让，包括搭建国际合作平台、建立信息分享机制、发挥政府间海洋学委员会等相关国际组织的作用。

第四，中国政府支持非洲集团提出的能力建设须是“有意义的”，倡导既要“授人以鱼”，更要“授人以渔”，通过教育、技术培训、联合研究等方式，切实提升发展中国家在养护和可持续利用 BBNJ 方面的内生能力。

六、跨领域问题

中国政府认为，讨论跨领域问题的关键是遵循《公约》的规定和精神，切实维护《公约》确立的国际海洋法律秩序，在《公约》所赋予的各项权利和义务之间保持合理平衡。

1995 年《鱼类种群协定》第 4 条规定：“本协定的任何规定均不应妨害《公约》所规定的国家权利、管辖权和义务。本协定应参照《公约》的内容并以符合《公约》的方式予以解释和适用。”中国政府认为，该规定为确定新国际文书与《公约》的关系提供了指引。

中国政府注意到，有关国家提出就新国际文书建立缔约国会议等履约机制和国际机构等问题。中国政府认为，新国际文书可建立缔约国会议就其履行情况进行审议。至于新国际文书是否需要建立以及建立何种国际机构，中国政府认为，这些问题应视新国际文书对海洋遗传资源、划区管理工具和环境影响评价的制度安排以及对能力建设和海洋技术转让作何规定，由各方在协商一致的基础上加以确定。对于新国际文书的签署、批准、加入、退出以及修订等最后条款问题，中国政府认为目前讨论此问题为时过早，此问题可在谈判新国际文书草案时再予以明确，1995 年《鱼类种群协定》有关规定可作为参考。

关于新国际文书有关争端解决的制度安排，考虑到 BBNJ 养护和可持续利用问题有其特殊性和专业性，不宜照搬《公约》第十五部分有关争端解决的规定。中国政府认为，如发生有关新国际文书解释或适用的争端，应优先由当事方通过谈判协商解决；如争端无法通过谈判协商得到解决，可考虑诉诸当事方明示同意的第三方程序。

The Government of the People's Republic of China on Elements of a Draft Text of an International Legally Binding Instrument under the United Nations Convention on the Law of the Sea on the Conservation and Sustainable Use of BBNJ

The Government of the People's Republic of China (the Chinese Government) attaches great importance to the conservation and sustainable use of marine biological diversity of areas beyond national jurisdiction (BBNJ). China has sent delegations to and actively participated in all the previous meetings of the *Ad Hoc* Open-ended Informal Working Group of BBNJ and the previous three sessions of the BBNJ Preparatory Committee (PrepCom). China is willing to continue playing a constructive role in the future BBNJ consultations and to make its contribution to the endeavor of the international community to better address the conservation and sustainable use of BBNJ.

The Chinese Government supports the statements made by the Group of 77 and China during the previous three BBNJ PrepCom sessions and the views expressed in the written submission of the G77 and China relating to BBNJ issues. Besides the aforementioned statements and views, the Chinese Government hereby would like to make some additional comments in its own capacity. It is noted that this Submission is without prejudice to the Chinese Government's possible further comments or proposals in future discussions.

I. General Position

The Chinese Government supports Resolution 69/292 entitled "Development of an international legally binding instrument under the United Nations Convention on the Law of the Sea on the conservation and sustainable use of marine biological diversity of areas beyond national jurisdiction" as adopted by the General Assembly of the United Nations (UNGA), highlighting that the mandate of the PrepCom is to "make substantive recommendations to the General Assembly on the elements of a draft text of an international legally binding instrument under the Convention". The relevant work should be carried out by parties concerned strictly as authorized by the Resolution. The final substantive recommendations to be made by the PrepCom should, to the extent possible, be based on consensus and reflect the common understanding of all parties. Meanwhile, the Chinese Government supports the PrepCom in addressing, together and as a whole, the topics identified in the 2011 package agreed to by all sides, namely marine genetic resources (MGRs), including sharing of benefits, area-based management tools (ABMTs), including marine protected areas (MPAs), environmental impact assessments (EIAs), capacity-building and the transfer of marine technology.

Firstly, the new international instrument, as a legally binding document under the framework of the United Nations Convention on the Law of the Sea (UNCLOS), should be consistent with the object and purpose of the UNCLOS, playing a supplementary and complementary role. It should not deviate from the principles and spirit of the UNCLOS or undermine its existing framework. Nor should it impair the integrity of the UNCLOS and the delicate balance therein. The freedoms and rights in respect of navigation, scientific research and fishing enjoyed by States under the UNCLOS should not be derogated. The provisions of the UNCLOS concerning the rights and obligations of the coastal States, including those concerning the rights and obligations over the continental shelf beyond 200 nautical miles, should not be affected.

Secondly, the new international instrument should not contravene current international law and existing global, regional or sectoral marine mechanisms. Nor should it undermine existing relevant legal instruments and frameworks and relevant global, regional and sectoral bodies. In particular, it should refrain from interfering with the mandates of bodies such as the Food and Agriculture Organization of the United Nations (FAO), Regional Fisheries Management Organizations and Arrangements (RFMO/As), the International Maritime Organization (IMO) and the International Seabed Authority (ISA). The new international instrument should facilitate cooperation and coordination with existing relevant international bodies, and avoid overlap or conflict of functions.

Thirdly, the relevant institutional arrangements of the new international instrument should have sound legal basis and solid scientific evidences, and maintain a reasonable balance between the conservation of BBNJ and its sustainable use.

Fourthly, the new international instrument should accommodate the interests and concerns of all sides. It should also base itself on the interests and needs of the international community as a whole and the absolute majority of States, especially those of the developing States. It should not overburden States, developing States in particular, by adding obligations and responsibilities beyond their capacity.

II. Marine Genetic Resources, including Questions on the Sharing of Benefits

MGRs in areas beyond national jurisdiction (ABNJ) are of tremendous actual or potential value for humankind. The institutional arrangements of the new international instrument should help to promote scientific research, encourage innovation, facilitate fair and equitable sharing of the benefits from the conservation and sustainable use of marine biological diversity of ABNJ, with a view to advancing the common well-being of humankind.

Definition

The Chinese Government considers that the definitions of genetic resource contained in the Convention on Biological Diversity (CBD) and the International Treaty on Plant Genetic Resources for Food and Agriculture (ITPGRFA) may serve as references for discussion on the definition of MGRs in the new international instrument. The definition in the new international instrument should include the following four elements: (1) animal, plant, microbe or other origin in the oceans and seas; (2) genetic materials containing functional units of heredity; (3) the actual or potential value; (4) the resources derived from ABNJ.

The Chinese Government is aware that the definitions of genetic resources contained in the CBD and the ITPGRFA do not encompass derivatives, which are products of biochemical synthesis without functional units of heredity. The definition of genetic resources in the new international instrument should not include derivatives, either.

The Chinese Government considers it necessary to make a

distinction with respect to fish in ABNJ. For fish as commodity, the UNCLOS, the Agreement for the Implementation of the Provisions of the United Nations Convention on the Law of the Sea of 10 December 1982 relating to the Conservation and Management of Straddling Fish Stocks and Highly Migratory Fish Stocks (UNFSA) and other international treaties already have detailed provisions, and the current RFMO/As have covered almost all the areas of the high seas. Thus the conservation and sustainable use of fishery resources should continue to be governed by existing RFMO/As and should not be regulated by the new international instrument. For fish as carrier of MGRs, the provisions concerning the freedom of marine scientific research in the UNCLOS should apply.

Access

During the discussions of the PrepCom, some States argued that access to MGRs consists of three types, namely*in situ* collection, *ex situ* collection and *in silico* analysis. The *in situ* collection means obtaining or collecting MGRs in natural environments in ABNJ. The *ex situ* collection and the *in silico* analysis refer to the collection of resources, information, materials and data resulting from the laboratory separation, identification, selection, cultivation and computer-based simulation analysis of MGRs samples obtained from *in situ* collection. The Chinese Government is of the view that the *in situ* collection in essence falls within the scope of scientific research in ABNJ as stipulated by the UNCLOS, and therefore free access should apply so as to facilitate the exploitation and sustainable use of MGRs.

Benefit Sharing

The collection of samples of MGRs in ABNJ, subsequent research and development in this regard, as well as the commercialization of useful products are characterized by demands for, *inter alia*, high technology, long periods of time, large investments and uncertainty in terms of outcomes, *etc.* The benefit sharing arrangements for MGRs should in general serve to promote the conservation and sustainable use of BBNJ, encourage marine scientific research and facilitate benefit sharing of MGRs for all humankind. The Chinese Government is of the opinion that on the premise of fully accommodating the concerns and needs of developing States, the PrepCom should give priority to non-monetary benefit sharing mechanisms such as easy access to samples, information exchange, transfer of technology and capacity-building. Meanwhile, The Chinese Government is open to discuss and explore the establishment of a monetary benefit sharing mechanism.

III. Area-Based Management Tools, including Marine Protected Areas

The Chinese Government supports the promotion of the conservation and sustainable use of BBNJ and attaches great importance to ABMTs, including MPAs.

Definition

ABMTs include various forms and approaches of management in addition to MPAs. The Chinese Government maintains that the definition of ABMTs to be regulated by the new international instrument should include but is not limited to the following three basic elements. (1) The objective: ABMTs should be aimed at the conservation and sustainable use of marine biological diversity. (2) The geographic scope: ABMTs should be applied only to areas in the high seas and the international seabed area. (3) The function: ABMTs should include different functions and management approaches.

With regard to the definition of MPAs, the Chinese Government suggests that the relevant definitions in the CBD and some regional instruments may serve as references to define MPAs in the new international instrument.

Principles and Approaches for Establishing Marine Protected Areas

The preamble of the UNCLOS clearly states that the States Parties are "[c] onscious that the problems of ocean space are closely interrelated and needs to be considered as a whole." In line with this spirit, when it comes to the principles and approaches for establishing MPAs, the integrated marine management approach could be considered in order to make up the shortfalls in the current regional and sectoral management approaches.

With regard to the specific guiding principles for establishing MPAs to be developed in the new international instrument, the Chinese Government in general associates itself with the statements on this issue made by G77 and China at the third session of the PrepCom and additionally emphasizes the following principles. (1) The principle of necessity: MPAs are tools rather than objectives, so they should be established on the premise of necessity. (2) The principle of proportionality: Conservation measures must be proportional to the objectives and effects of conservation and should be applied in a cost-effective manner. (3) The principle of scientific evidence: The establishment of MPAs needs to be based on solid scientific evidence, and to evaluate the potential threats to and risks for the ecosystems, habitats and populations to be protected. (4) The principle of different levels of protection: Different management tools should be applied according to the respective characteristics of different sea waters, ecosystems, habitats and populations. (5) The principle of international cooperation: All States and international organizations are obliged to collaborate with each other in the establishment of MPAs.

The Chinese Government deems that the establishment of MPAs needs to meet certain substantive and procedural requirements. With respect to substantive requirements, the establishment of MPAs should have, *inter alia*, clear conservation objectives, certain conservation targets, specific protection scope, appropriate protection measures and reasonable time limit, *etc*. As for procedural requirements, the establishment of MPAs should follow specific procedures, including submission, consultation, review, decision-making, management, monitoring and surveillance.

In accordance with Resolution 69/292 of the UNGA, the new international instrument, as an implementing agreement of the UNCLOS, should aim at identifying lacuna and filling gaps. Whatever kind of institutional arrangement will be made in the new international instrument, it should not prejudice the functions and mandates of existing institutions, such as ISA, IMO and RFMOs, or undermine the relevant provisions in existing international treaties. However, in the very process of establishing MPAs, it is necessary for the new international instrument to cooperate and coordinate with existing regional and sectoral institutions.

The relationship between provisions concerning ABMTs including MPAs in existing mechanisms and those to be made in the new international instrument involves the relationship between existing international treaties and the new international instrument, which should be carefully and properly addressed in line with the general principles concerning treaty application as provided for under the Vienna Convention on the Law of Treaties.

IV. Environmental Impact Assessments

The Chinese Government attaches great importance to marine EIAs, and believes that the institutional arrangements in respect of EIAs in the new international instrument should comply with the basic legal framework and the procedural elements provided by the UNCLOS. It should also take into account the provisions in respect of EIAs in other international instruments.

Article 206 of the UNCLOS provides that "[w]hen States have reasonable grounds for believing that planned activities under their jurisdiction or control may cause substantial pollution of or significant and harmful changes to the marine environment, they shall, as far as practicable, assess the potential effects of such activities on the marine environment." In accordance with this Article, the Chinese Government believes that the subject to conduct EIAs under the new international instrument should be States planning to undertake marine activities. The object of EIAs should be the planned "activities" under the jurisdiction or control of States, excluding strategic environment assessments (SEAs). The threshold to trigger EIAs is to "have reasonable ground for believing" that such activities "may cause substantial pollution of or significant and harmful changes to the marine environment". In addition, in the light of the factors such as scientific information, technical methods, cost and capacity, it should be carefully deliberated whether assessment of cumulative impacts is "as far as practicable".

Article 194 of the UNCLOS specifies that States shall ensure that "pollution arising from incidents or activities under their jurisdiction or control does not spread beyond the areas where they exercise sovereign rights in accordance with this Convention." The Chinese Government believes that since the UNCLOS has made provisions for trans-boundary impacts of activities within the areas of national jurisdiction, the scope of EIAs under the new international instrument should be limited to activities in the ABNJ, which include activities that *may cause significant environmental impacts to* the areas within the jurisdiction of the coastal States, but exclude activities that *take place in* such areas.

In accordance with Resolution 69/292 of the UNGA, the new international instrument, as an implementing agreement of the UNCLOS, should not undermine existing relevant legal instruments and frameworks and relevant global, regional and sectoral bodies. The provisions concerning EIAs in the new international instrument should not prejudice the functions and mandates of existing international institutions and relevant provisions in existing international instruments. The Chinese Government is of the opinion that existing rules governing EIAs, *e. g.* those applicable to deep sea fisheries in the high seas, deep sea mineral exploration in the international seabed area and waste dumping, should not be undermined.

The Chinese Government notes that some States propose to produce a list of activities subject to EIAs. The Chinese Government considers that marine environmental impacts vary from one type of activities to another, and the listing approach has certain limitations in this regard. If all parties agree to conceive such a list, the Chinese Government suggests that the list should be non-exhaustive and advisory by nature, with no legally-binding force.

V. Capacity-Building and Transfer of Marine Technology

Capacity-building and transfer of marine technology are important means to improve the capacity of developing countries in the conservation and sustainable use of BBNJ, and remain indispensable for realizing the overall objective of marine environmental protection and sustainable development. The Chinese Government is supportive of the general position of G77 and China on capacity-building and transfer of marine technology, and wishes to make the following additional observations:

Firstly, the provisions concerning capacity-building and transfer of marine technology should be based on relevant provisions in Part XIV of the UNCLOS, following the principles of pertinence, effectiveness, equality and mutual benefit, and win-win cooperation.

Secondly, the new international instrument should take full account of the needs and interests of developing countries, in particular small islands developing States, the least developed countries, landlocked and geographically disadvantaged States as well as countries with special interests and needs.

Thirdly, the new international instrument should encourage various forms of international cooperation, including the creation of an international cooperation platform, the establishment of information sharing mechanisms as well as the utilization of the Intergovernmental Oceanographic Commission (IOC) and other relevant international organizations, with a view to strengthening the capacity-building of and the transfer of marine technology to the developing countries.

Fourthly, the Chinese Government concurs with the proposal by the African Group that capacity-building should be "meaningful". The Chinese Government advocates that while it is important "to give fish", it is more important to "teach how to fish", which means that through approaches such as education, technical training and joint research, the endogenous capacity of developing countries for the conservation and sustainable use of BBNJ could be effectively improved.

VI. Cross-Cutting Issues

The Chinese Government is of the view that the key to discussing cross-cutting issues is to follow the provisions and the spirit of the UNCLOS, while upholding the established international maritime legal order and maintaining a reasonable balance between the rights and obligations conferred by the UNCLOS.

Article 4 of the UNFSA stipulates that "Nothing in this Agreement should prejudice the rights, jurisdiction and duties of States under the Convention. This Agreement should be interpreted and applied in the context of and in a manner consistent with the Convention". It is the view of the Chinese Government that this Article does provide some guidance for addressing the relationship between the new international instrument and the UNCLOS.

The Chinese Government takes note of some States' proposal to establish Conference of Parties (COP) as a review mechanism and to establish international bodies under the new international instrument. The Chinese Government is of the view that the COP may be established for the purpose of reviewing the implementation of the new international instrument. Nevertheless, with regard to whether, and what types of, international bodies should be established, the Chinese Government is of the view that these issues hinge on the contents of the institutional arrangements for MGRs, ABMTs and EIAs, and the provisions concerning capacity-building and transfer of marine technology in the new international instrument, and should be addressed by consensus. With respect to the final provisions of the new international instrument, including those concerning signature, ratification, accession, withdrawal and amendment, the Chinese Government considers it premature to address these issues at the present stage. These issues may be clarified in the course of negotiating the draft text of the new international instrument, and the relevant provisions of the UNFSA may serve as references.

Regarding the institutional arrangements for dispute settlement in the new international instrument, taking into account the special characteristics of the conservation and sustainable use of BBNJ and the expertise required to address the issue, it is inappropriate to directly apply the dispute settlement provisions contained in Part XV of the UNCLOS. The Chinese Government is of the view that when a dispute concerning the interpretation or application of the new international instrument arises, the parties concerned should first resort to negotiation and consultation to settle the dispute. If the dispute is not settled thereafter, the parties may consider submitting the case to a third-party procedure based on explicit mutual agreement.

BBNJ 国际协定谈判预备委员会
第一次会议中国代表团发言汇编

国家管辖范围以外区域海洋生物多样性（BBNJ）养护和可持续利用问题国际协定谈判预备委员会第一次会议于 2016 年 3 月 28 日在纽约联合国总部举行。会上，中国代表团建设性参与讨论。现将代表团在预委会第一次会议上的主要发言内容摘录如下，以备工作参考。

一、一般性发言

中国代表团支持预委会以 2011 年各方达成的共识为基础，“一揽子”同步推进海洋基因资源、划区管理工具、环境影响评估、能力建设和海洋技术转让等问题，并对制定 BBNJ 国际文书提出四点看法：

第一，新国际文书不能与现行国际法以及现有全球性、区域性和专门性的海洋机制相抵触。新国际文书是在《联合国海洋法公约》（《公约》）框架下制定的国际法律文件，应是对《公约》的补充和完善，不能偏离《公约》的原则和精神，不能损害《公约》建立的制度框架，不能损害《公约》的完整性和微妙平衡。

第二，新国际文书应兼顾各方利益和关切，立足于国际社会整体和绝大多数国家的利益和需求，特别是应顾及广大发展中国家的利益。预委会最终就实质要素所提出的建议，应尽最大努力在协商一致的基础上反映各方共识。

第三，新国际文书的有关制度设计和安排应在海洋环境保护与可持续利用之间保持合理平衡。

第四，新国际文书的有关制度设计和安排应有充分的法律依据、坚实的科学基础，并符合客观实际需要。

二、关于海洋保护区等划区管理工具

海洋保护区等划区管理工具是 2011 年一揽子共识中的重要要素。中国代表团支持泰国代表团代表 77 国集团和中国就该问题所作发言并提出五点看法：

第一，养护和可持续利用是新协定的两大目标，设立海洋保护区等划区管理工具应保持养护和可持续利用的合理平衡，不能厚此薄彼。

第二，设立海洋保护区等划区管理工具应符合《公约》的目的和宗旨，不能影响各国依照《公约》享有的公海自由和权利，不能影响沿海国依据《公约》享有的对 200 海里外大陆架的主权权利。

第三，设立海洋保护区等划区管理工具应以养护的实际需要为前提，并应具备坚实的科学基础。保护区的设立既要建立一般门槛和标准，也要考虑不同海域的具体情况，不能搞一刀切，应具体海域具体分析。同时，保护区的范围和保护的具体措施应与实际需求相适应。

第四，应在《公约》框架下为海洋保护区等划区管理工具作出制度性安排，包括对保护区设立的申请和审查，以及对保护区的管理、监督、期限、撤销等作出规定。

第五，海洋保护区等划区管理工具的设立和管理应与现有全球、区域和专门的制度和机制相协调。

三、关于环境影响评估

中国代表团支持泰国代表团代表 77 国集团和中国就环境影响评估所作的发言。环境影响评估是保护海洋生物多样性的重要手段，但有关评估内容、标准和方法等并不确定。中国代表团就该问题提出以下几点看法：

第一，环境影响评估的制度设计应符合《公约》的规定。根据《公约》第 206 条，只有在相关活动可能对海洋环境造成“重大污染或重大和有害的变化”的情况下，才能就这种活动对海洋环境的可能影响作出评估。

第二，环境影响评估是保护和保全海洋环境的预防性措施，有关制度安排不仅应有利于促进海洋环保，而且应符合可持续利用的目标。

第三，应明确评估的区域和范围。评估的范围应主要限于国家管辖范围以外的公海和国际海底区域发生的活动。环境影响评估应不妨碍各国依据《公约》在公海和“区域”享有的自由和权利；应充分顾及沿海国依据《公约》享有的主权权利和管辖权，非经沿海国同意，第三方不得对国家管辖范围内的海域进行环境影响评估。

第四，应建立一套完整的评估程序。应明确如何启动环境影响评估，在评估过程中应运用最佳科学数据，符合国际最佳做法，评估程序应公开、透明和包容，并照顾所有利益攸关方的利益和关切，同时还应加强对环评的管理和监督。

第五，新协定中环境影响评估的制度设计应充分借鉴现有国际机构制定的有关规则，并与之加强协调。

此外，中国代表团就环境影响评估问题补充一点意见：

海洋环境影响评估是保护和保全海洋环境的预防性措施，有关制度安排应同时兼顾海洋保护和可持续利用两个目标。在海洋环保方面，根据《公约》相关规定，只有在相关活动可能对海洋环境造成“重大污染或重大和有害的变化”的情况下，才能对这种活动对海洋环境的可能影响作出评估。在可持续利用方面，环境影响评估应从国际社会的共同利益出发，不应妨碍各国根据《公约》享有的权利和自由，包括充分考虑各国依据公约享有的科研自由和航行自由，以及长期可持续合理利用海洋资源的权利。

四、关于能力建设和技术转让

中国代表团支持泰国代表团代表 77 国集团和中国就能力建设和技术转让所作的发言，提出了以下看法：

第一，新协定应充分照顾发展中国家的需要和利益。能力建设和技术转让是提升发展中国家养护和可持续利用海洋生物多样性能力的重要手段，也是实现海洋环境保护和可持续发展整体目标不可或缺的重要方面。

第二，新协定应就能力建设和技术转让作出制度性规定，以全面落实《公约》第十四部分规定的有关义务，促进 2030 可持续发展议程的全面实施。

第三，新协定应鼓励通过多种形式的国际合作加强发展中国家能力建设，特别是照顾小岛屿发展中国家、最不发达国家以及有特殊利益需求的国家的利益和关切，切实使包括这些国家在内的所有国家从海洋生物多样性的养护和可持续利用中获益。

五、下步工作

中国代表团支持泰国代表 77 国集团和中国所作发言，并就

BBNJ 新协定和预委会的下步工作重申了以下看法：

第一，BBNJ 新协定应旨在实施《公约》第十二、十三、十四部分等的精神和内容，而不是根本改变《公约》的基本原则和固有平衡。新协定是对《公约》的补充，应按照《公约》的精神和原则填补有关具体规则的法律空白，重在编纂现有国际实践，并在此基础上逐渐发展。

第二，根据联大第 69/292 决议，BBNJ 预委会的职权应在 2011 年一揽子共识的基础上，就新协定的实质要素提出建议。在目前阶段，预委会应重点就实质要素的各个方面，包括范围问题广泛听取各方意见，就有关问题展开充分讨论，以不断积累和凝聚共识。

第三，根据联大 69/292 号决议，新协定不能损害现有国际法律文书和框架以及全球性、区域性和专门性机构。但现有国际法律文书和框架的范围和具体内容是什么，目前尚未有统一认识。中国代表团认为应对此问题予以明确，必要时可考虑邀请相关国际组织作技术报告，厘清新协定与现有国际机构和规则的关系。这将有助于各代表团全面了解有关海洋保护区与环境影响评估的“现有法”（Lex Lata），深化我们对 BBNJ“应有法”（Lex Ferenda）的讨论。

Statement by the Chinese delegation on the first session of the Preparatory Committee on the negotiation of an international instrument on BBNJ

1. General Statements

My delegation supports the preparatory committee in working on the basis of the common understanding reached in 2011 and pushing forward in a synchronized manner work on issues of marine genetic resources, area-based management tools, environmental impact assessments and the transfer of marine technology. The Chinese delegation wishes to make four observations regarding the negotiation and formulation of an international instrument on BBNJ.

First, the new international instrument must not contradict contemporary international law and existing global, regional and specialized marine mechanisms. The new instrument is an international legal document under the framework of the UN Convention on Law of the Sea. As such, it should be a complement to and improvement of the Convention and must not deviate from the principles and spirit of the

Convention, undermine the institutional framework established by the Convention, or affect the integrity and delicate balance of the Convention.

Secondly, the new international instrument should base itself on the interests and needs of the entire international community and the absolute majority of states, and accommodate the interests and concerns of all sides, particularly those of the developing countries. The recommendations to be put forward by the PrepCom on substantive elements should be based on consensus to the extent possible so as to reflect the common understanding of all sides.

Thirdly, the institutional design and arrangement stipulated by the new international instrument should strike a reasonable balance between marine environmental protection and sustainable utilization.

Fourthly, such institutional design and arrangement should have a sound legal and scientific basis and meet practical needs.

2. Consideration of measures such as area-based management tools, including marine protected areas

China associates itself with the statement delivered by the representative of Thailand on behalf of the Group of 77 and China and wishes to make five points on area-based management tools, including MPA.

First, conservation and sustainable utilization are the two major goals of the new agreement and in establishing area-based management tools such as the MPA, a reasonable balance between these two goals should be striken instead of focusing on one while neglecting the other.

Secondly, the establishment of MPA and other area-based management tools should be in line with the purposes and principles of UN-

CLOS. It should not affect countries' exercise, according to the Convention, of their rights and freedoms on the high seas, nor should it affect the sovereign right of coastal states over their outer continental shelves beyond 200 nautical miles.

Thirdly, the establishment of MPA and other area-based management tools should be premised on practical needs for conservation and based on a sound scientific basis. While it is necessary to set up general thresholds and criteria for the establishment of the protected area, consideration should be given to the conditions of specific marine areas and analysis made on that basis instead of using a one-size-fits-all approach. Meanwhile, the scope of the MPA and concrete protective measures should correspond to practical needs.

Fourthly, institutional arrangements should be made within the framework of UNCLOS for MPA and other area-based management tools, including rules for the application for the establishment of MPAs and review of such applications, and for the management, monitoring, timeframe and cancellation of MPAs.

Fifthly, the establishment and regulation of MPA and other area-based management tools should be coordinated with the existing global, regional and sectoral institutions and mechanisms.

3. Environmental Impact Assessment

The Chinese Delegation associates itself with the statement by the representative of Thailand on behalf of G-77 and China on EIA. EIA is an important means to protect marine biodiversity. However, the elements, criterion and methods of such assessment are still uncertain. My delegation wishes to put forward the following observations.

First, the EIA regime should be consistent with the provisions of the Convention. According to article 206 of the Convention, EIA should be conducted only when the relevant activities might cause serious pollution or significant and harmful change to the marine environment.

Second, EIA is a precautionary measure to protect and preserve marine environment. The relevant regime or arrangements should not only contribute to marine protection, but also in keeping with the goal of sustainable use of marine resources.

Third, the areas and the scope of the assessment should be clearly defined. The scope of assessment should be limited to the high seas and activities that occur on international seabed beyond national jurisdiction. EIA should not compromise the freedom and interests enjoyed by all countries on the high seas and the Area under the Convention. The sovereign rights and jurisdiction of the coastal states under the Convention should be fully taken into account. A third party can not conduct EIA in areas within national jurisdiction.

Fourth, a set of comprehensive assessment procedures should be established. There should be clear provision on how to activate EIA, in which the best available scientific data should be used and the work should be in keeping with international best practices. The assessment process should be open, transparent and inclusive, which takes care of the interests and concerns of the stakeholders while ensuring enhanced administration and monitoring.

Fifth, in designing the EIA regime for the new agreement, we must consult the existing rules of the relevant international organizations with which we should enhance coordination.

4. Capacity Building and Transfer of Technology

The Chinese delegation associates itself with the statement made by the representative of the delegation of Thailand on behalf of the Group of 77 and China on capacity building and transfer of technology and wishes to make the following additional remarks.

First, the new instrument should fully consider the needs and interests of the developing countries. Capacity building and transfer of technology are among the important means to scale up the capacity of the developing countries for the preservation and sustainable use of marine biological diversity as well as important integral aspects of the efforts to realize the overall objective of marine environmental protection and sustainable development.

Secondly, the new instrument should set forth statutory regulations on capacity building and transfer of technology in order to comprehensively fulfill all the relevant obligations under Part XIV of the UNCLOS and promote the full implementation of the Agenda 2030 for sustainable development.

Finally, the new instrument should encourage the efforts to strengthen the capacity building of the developing countries through various forms of international cooperation and, in particular, take account of the interests and concerns of the small island developing states, the least developed countries and those countries that have special interests and needs with a view to effectively ensuring that all countries, including the above - mentioned countries will benefit from the preservation and sustainable use of marine biological diversity.

5. The work of the Preparatory Committee in the next phase

The Chinese delegation associates itself with the statement made by the representative of Thailand on behalf of the G-77 and China and would like to make the additional remarks regarding the new instrument on BBNJ and the work of the Preparatory Committee in the next phase.

First of all, the new instrument on BBNJ should be aimed at implementation of the letter and the spirit of the relevant provisions ofparts XII, XIII, and XIV of the Convention, rather than altering the fundamental principle and inherent balance of the Convention. As a complement to the Convention, the new instrument should serve to fill the legal gap in the relevant provisions in accordance with the spirit and principle of the Convention with the the focus on codifying and further developing current international practices in this field.

Secondly, the task of the Preparatory Committee on BBNJ, as mandated in resolution 69/292 of the General Assembly, is to make recommendations on the substantive elements of the new instrument on the basis of the 2011 package. At present, the focus of the Preparatory Committee should be broad exchange of views on various aspects of the substantive elements including the question of the scope and thorough deliberation of the relevant issues with a view to building and reaching consensus on the matter.

Thirdly, according to resolution 69/292 of the General Assembly, the new instrument should notundermine existing international legal instruments and frameworks as well as global, regional and specialized mechanisms. However, currently there is no common understanding on the scope and specific content of the existing international legal instruments, frameworks and international mechanisms. In the view of the Chinese delegation, this matter needs to be clarified and relevant international organizations should be requested , as appropriate, to present a technical report in order to lay out the relationship between the new instrument and existing international mechanisms and rules. This will help delegations obtain a comprehensive understanding of the Lex Lata regarding MPA and EIA and conduct in-depth discussion of the Lex Ferenda regarding BBNJ.

BBNJ 国际协定谈判预备委员会
第二次会议中国代表团发言汇编

国家管辖范围以外区域海洋生物多样性（BBNJ）养护和可持续利用问题国际协定谈判预备委员会第二次会议于 2016 年 8 月 26 日在纽约联合国总部举行。会上，中国代表团发言有理有据，在多个关键问题上发挥定调和引领作用。现将代表团在预委会第二次会议上的主要发言内容摘录如下，以备工作参考。

一、关于海洋基因资源

泰国代表团代表 77 国集团加中国就本议题作了发言，中国代表团支持上述发言，并就海洋基因资源（MGR）的定义发表以下看法。

MGR 的定义关乎其获取和惠益分享的范围和制度安排。《生物多样性公约》（CBD）对基因资源和基因材料作出了界定，为我们定义 MGR 提供了必要参考，但其主要是针对国家管辖范围内的基因资源，其与国家管辖范围外区域的基因资源在种类和特性等方面不尽一致，新协定不宜直接照搬上述定义，应根据国家管辖范围以外的实际情况进行适应化调整。

关于海洋基因资源的定义是否应包括衍生物，目前尚未形成普遍共识。CBD 中的“基因资源”定义仅包括遗传物质本身，未包括衍生物。尽管《名古屋议定书》对“利用遗传资源”和“衍生物”作出界定，但各国对“衍生物”是否属于基因资源的认识不同。世界知识产权组织的“知识产权与遗传资源、传统知识和民间文学艺术政府间委员会”就与基因资源相关的知识产权问题进行反复磋商，各方仍未对“利用遗传资源”“衍生物”等定义达成一致。

中国代表团还认为，BBNJ 国际文书中关于海洋基因资源的定义包括以下四个核心要素：

一是来自海洋的动物、植物和微生物或其他来源；

二是含有遗传功能单位的遗传材料；

三是具有实际或潜在价值；

四是地理范围是国家管辖范围以外区域。

此外，BBNJ 国际文书中海洋基因资源的定义不应包括衍生物。

二、关于划区管理工具

中国代表团支持泰国代表团代表 77 国集团加中国就本议题所作发言，并以国家立场就划区管理工具的定义作简要补充。

划区管理工具包括一系列不同管理方法，在实践中形式多种多样，现阶段对其作出定义有一定难度。中国代表团初步认为，划区管理工具定义包括但不限于以下三个基本要素：

一是目标要素。划区管理工具应以养护和可持续利用海洋生物多样性为目标。

二是地理范围要素。划区管理工具适用的地理范围应是公海和国际海底区域的特定区域。

三是功能和手段要素。划区管理工具应包括不同功能和管理方法。

三、关于环境影响评估

中国代表团支持泰国代表团代表 77 国集团加中国就本议题所作发言，并就环境影响评价的依据提出以下看法：

BBNJ 国际文书是在《联合国海洋法公约》（《公约》）框架下制定的法律文件，应是对《公约》的实施。《公约》第 206 条对国家管辖范围以外区域的活动进行环境影响评价提供了法律框架。中国代表团认为，新文书有关环境影响评价制度应遵循《公约》第 206 条的框架，主要包括以下要素：

（一）评价的主体是国家。根据《公约》规定，环境影响评价由各国自主进行。

（二）评价的对象是发生在国家管辖范围以外区域的“活动”（activities）。原则上，我们理解战略环评不在此内。

（三）评价的启动门槛是“有合理依据认为”（reasonable grounds for believing），“可能造成重大污染或重大和有害的变化”（may cause substantial pollution of or significant and harmful changes）。只有在相关活动可能造成“重大污染或重大和有害的变化”的情况下，才需要进行环境影响评价。

（四）评价具有可操作性。环境影响评价应“在实际可行范围内”（as far as practicable）进行，意味着有关评价需在实体和程序上都具有可操作性。

（五）评价的内容是有关活动“对海洋环境的可能影响”（the potential effects of such activities on the marine environment）。在 BBNJ 语境下，各国需就有关活动对国家管辖范围以外区域的海洋生物多样性的可能影响作出评价。

（六）评价结果的报告。根据《公约》规定，各国应发表有关评价报告，并向主管的国际组织提供。中国代表团认为，上述环境影响评价的要素是 BBNJ 新文书相关规定的依据，希望各方积极考虑。

四、关于能力建设和技术转移

中国代表团支持泰国代表团代表 77 国集团加中国就本议题所作发言，并补充以下看法：

第一，新协定在能力建设和技术转让方面应遵循针对性、有效性、平等互利、合作共赢等原则。

第二，新协定应着力加强信息和技术的便利获取和分享，积极探讨建立 BBNJ 各相关领域全面的信息分享机制，充分利用政府间海洋学委员会（IOC）、海洋生物地理信息系统（OBIS）等已有的国际信息交换平台。

第三，有关制度安排应充分考虑发展中国家的利益和实际需求，特别是照顾小岛屿发展中国家、最不发达国家、内陆国和地理不利国以及有特殊利益需求国家的利益和关切。还应特别考虑不同国家在不同发展阶段的切实需要。我们赞同非洲集团所提出的相关能力建设需是“有意义的”。

第四，针对发展中国家开展的能力建设项目，既要“授人以鱼”，更要“授人以渔”，通过教育、科技培训、联合研究等方式，切实提升发展中国家在养护和可持续利用 BBNJ 方面的内生能力。

第五，新协定应鼓励通过多样形式的国际合作加强能力建设和技术转让，努力搭建国际合作的平台，充分发挥相关国际组织的作用。

五、关于跨领域问题

中国代表团支持泰国代表团代表 77 国集团加中国就本议题所

作发言，并提出以下看法：

一、BBNJ 国际文书是在《公约》框架下制定的法律文件，应是对《公约》的补充和完善，不能偏离《公约》的原则和精神，不能损害《公约》建立的制度框架，不能损害《公约》的完整性和微妙平衡。各国根据《公约》享有的航行、科研、捕鱼等方面的权利和义务不应受到减损。沿海国依据《公约》享有的权利和义务，包括对 200 海里以外大陆架的权利和义务，不应受到减损。

二、预委会相关工作应严格遵循联大第 69/292 号决议授权。未来的 BBNJ 国际文书不能损害现有相关法律文书或框架以及相关全球性、区域性和专门性机构，特别是不能干预联合国粮农组织、区域性渔业组织、国际海事组织、国际海底管理局等机构的职权，不能改变上述组织在各自框架下相关条约规定的权利和义务。

三、BBNJ 国际文书应致力于促进与现有相关国际机构和机制的协调与合作，避免有关工作或权限的重复或重叠。

六、关于海洋保护区

中国代表团认为，BBNJ 国际文书中海洋保护区涉及实体要素和程序要素两个方面。

一、实体要素可包含但不限于以下方面：

（一）必要性。海洋保护区的设立应以确有必要为前提。

（二）法律和科学依据。海洋保护区应依法设立，并根据科学证据原则，进行充分科学论证，以养护的实际需要为前提。

（三）保护对象和目标。海洋保护区的设立应根据不同海域的具体情况和特殊性，确定具体的保护对象和目标。

（四）保护范围。海洋保护区应有明确的地理界限和范围，并合理确定保护的面积。

（五）保护措施。相关保护措施应具体，并与保护的具体对象和目标相适应，具有可操作性。

（六）保护期限。海洋保护区应根据保护目标的需要设定合理的期限。CCAMLR 关于设立海洋保护区一般框架的养护措施（CM91-04）也明确规定了保护期限问题。

二、程序要素可包含但不限于以下方面：

（一）谁有权提议；

（二）谁有权对提案进行审查；

（三）谁有权对提案作出决定；

（四）谁应当执行，包括遵约、监测和审查等。

Statement by the Chinese delegation on the second session of the Preparatory Committee on the negotiation of an international instrument on BBNJ

1. Marine Genetic Resources

The Chinese delegation echoes the statement made by Thailand on behalf of the Group of 77 and China on this item and wishes to comment on the definition of the marine genetic resources (MGR).

MGR, by definition, relates to both the scope and institutional arrangement of its access and benefit sharing. The Convention on Biological diversity (CBD) provides for the definitions of genetic resources and genetic materials and serves as a necessary reference for our efforts to define MGR. However, these definitions are mainly applicable to genetic resources within national jurisdiction, which are not entirely consistent with genetic resources beyond national jurisdiction in either variety or attribute. Therefore, we caution against incorporating them directly into the new agreement and suggest that changes be made to adapt to specific realities beyond national jurisdiction.

As for whether the definition of MGR should include derivatives, no universal consensus has yet been reached. Genetic resources, as per CBD, only pertains to genetic resources per se to the exclusion of derivatives. It is true that the Nagoya Protocol does define "the utilization of genetic resources" and "derivatives", but countries vary in their perceptions as to whether "derivatives" should fall under genetic resources. Despite the series of consultations on intellectual property rights (IPR) that relate to genetic resources hosted by the WIPO Intergovernmental Committee on Intellectual Property and Genetic Resources, Traditional Knowledge and Folklore (IGC), no agreements were reached on the definitions of "the utilization of genetic resources" and "derivatives". We deem the above-mentioned issues in need of further discussions and welcome the inputs from the experts.

The Chinese delegation believes that a definition of Marine Genetic Resources (MGRs) in the international instrument on BBNJ should include the following four essential elements:

1) (are derived) / come from marine animals, plants, microbial and other origins;

2) are the genetic materials containing functional units of heredity;

3) are having actual or potential values; and

4) their geographical coverage should be the areas beyond national jurisdiction.

In addition, the Chinese delegation does not think that the definition of MGRs in the international instrument on BBNJ should include drivatives.

2. Area-based Management Tools

The Chinese delegation aligns itself with the statement made by the delegation of Thailand on behalf of G77 and China. From my national perspective, I would like to make a brief addition on the definition of area-based management tools.

China has noted that area-based management tools include a whole array of management methods which have evolved in practice into various forms and thus pose some difficulty for definition at this stage. In our preliminary view, the definition of area-based management tools includes, but are not limited to, the following three basic elements:

1) The objective element

The area-based management tools should be aimed at the conservation and sustainable use of marine biodiversity.

2) The geographic scope element

The applicable geographical scope for area-based management tools should be specially designated areas in the high seas and the international seabed.

3) The functions and means element

The area-based management tools should include different functions and methods of management.

3. EIA

The Chinese delegation associates itself with the statement made by Thailand on behalf of G77 and China. With regard to the element of the environmental impact assessment (EIA), we would like to make the following comments.

The BBNJ international instrument, a legal document being developed under the framework of the UNCLOS, should be the implementation of the Convention. Article 206 of the Convention provides the legal framework related to the EIA on activities in areas beyond the national jurisdiction. The Chinese delegation is of the view that the new instrument on EIA must conform to the framework under article 206 of the Convention and include the following elements:

1) The Subject of Assessment. States should be the principal actor in conducting the assessment. As stipulated by the Convention, EIA is to be independently carried out by states.

2) The Object of Assessment. The assessment should focus on the activities in areas beyond the national jurisdiction. Our understanding is that, in principle, strategic environmental assessment (SEA) is not included.

3) The Threshold of Assessment. The EIA is necessitated when there are "reasonable grounds for believing" that relevant activities "may cause substantial pollution of or significant and harmful changes to" the marine environment.

4) The Operability of Assessment. The EIA should proceed within the scope that is "as far as practicable". This means the relevant assessment should possess operability in terms of entities and processes.

5) The Content of Assessment. The "potential effects of (such) activities on the marine environment" should be the main thrust of the EIA. In the context of BBNJ, this requires the States to assess potential impact on marine biodiversity caused by relevant activities in areas beyond their national jurisdiction.

6) The Reports of the Assessment Result. In accordance with the Convention provisions, States should publish reports of the result of such assessment, copies of which are to be forwarded to the competent international organizations.

4. Capacity Building and Technology Transfer

The Chinese Delegation also aligns itself with the statement on this item made by the distinguished representative of Thailand, on behalf of the G77 and China. Now I would like to add the following comments:

1) The new agreement should be targeted and effective in the aspect of Capacity Building and Transfer of Technology, and adhere to the principle of Win-win cooperation on an equal footing with mutual benefits.

2) The new instrument should emphasize facilitating access to and sharing of information and technology, and proactively explore the establishment of a comprehensive information sharing mechanism covering all relevant aspects of BBNJ, while fully utilizing existing international platforms for information exchange such as the Intergovernmental Oceanographic Commission (IOC), and the Ocean Biogeographic Information System (OBIS).

3) The relevant arrangements should give due consideration to the interests and genuine needs of the Developing Countries, particularly the interests and concerns of Small Island developing countries, the LDCs, the landlocked and geographically disadvantaged states and those with special needs. Special consideration should also be given to the real needs of various countries at different stages of development. We support the idea of the African Group that the relevant capacity building needs to be "meaningful".

4) Capacity building projects targeting the developing countries should not only "give them fish", but also, more importantly, "teach them how to fish", so as to enhance the internal capacity of the Develo-

ping Countries in the Conservation and Sustainable use of BBNJ through such means as education, scientific and technological training and joint research.

5) The new instrument should encourage diversified forms of international cooperation for Capacity Building and Transfer of Marine Technology. Efforts should be made to establish international cooperation platforms while fully leveraging the role of relevant international organizations.

5. Cross-cutting Issues

The Chinese delegation associates itself with the statement by Thailand on behalf the Group of 77 and China under this agenda item and would like to make the following additional comments.

First, the international instrument on BBNJ, as a legally binding instrument developed under the framework of the United Nations Convention on the Law of the Sea (UNCLOS), should be a supplement to and a further improvement on UNCLOS. In no way should it deviate from the principles and spirit of the Convention, or undermine its framework. Nor should it impair the integrity of the Convention and upset its delicate balance. The rights and obligations of the States under UNCLOS in navigation, scientific research, fishing, among others, should be without prejudice to. The rights and obligations of coastal States over their continental shelf beyond 200 nautical miles should not be impaired or altered.

Second, the relevant work of the PrepCom should strictly abide by the mandate from the GA resolution 69/292. The future international instrument on BBNJ should not undermine other existing relevant legal instruments and frameworks as well as the terms of reference of the relevant global, regional and sectoral bodies, in particular, those of FAO, RFMOs, IMO and ISA should never be encroached upon. In no way may it change the rights and obligations of the above-mentioned organizations specified in relevant treaties in their respective frameworks.

Third, the international instrument on BBNJ should be aimed at promoting coordination and cooperation with existing relevant international organizations and mechanisms and avoid duplication or overlaps in their respective efforts and mandates.

6. Marine Protected Areas

The Chinese delegation believes that marine protected areas to be provided for in the international instrument on BBNJ involve both substantive and procedural elements.

First, the substantive elements could contain but not be limited to the following aspects:

i. Necessity. Marine protected areas should be established on the premise of material necessity.

ii. Legal and scientific basis. Marine protected areas should be established in accordance with law, comprehensively evaluated in a scientific way based on the principle of scientific evidence and premised on the need for conservation.

iii. The target of protection and its objective. Specific targets and objective of the protection should be set for a marine protected area in light of its actual conditions and special nature.

iv. Scope of protection. Marine protected areas should have clearly delineated geographic boundaries and scope and reasonably defined area of protection.

v. Measures of protection. The relevant protective measures should be fit for purpose, operational and appropriate to the specific targets and objective.

vi. Duration of protection. Reasonable timeframe should be set out for marine protected areas based on the needs of the objective of the protection activities.

Second, procedural elements could contain but not be limited to the following aspects:

i. Who has the right to make proposal;

ii. Who has the right to review the proposal;

iii. Who has the right to make a decision on the proposal;

iv. Who should implement, including compliance, monitoring and review mechanisms.

BBNJ国际协定谈判预备委员会第三次会议中国代表团发言汇编

国家管辖范围以外区域海洋生物多样性（BBNJ）养护和可持续利用问题国际协定谈判预备委员会第三次会议于2017年3月28日在纽约联合国总部举行。现将代表团在预委会第三次会议上的主要发言内容摘录如下，以备工作参考。

一、关于海洋遗传资源

中方支持由厄瓜多尔代表77国集团加中国所作的发言。在此，中方提出看法如下。

中方注意到在昨天和今天讨论中许多代表提到捕鱼问题，特别是要不要区分作为商品的鱼类和作为海洋遗传资源的鱼类。根据联大第69/292号决议，新国际文书“不应损害现有有关法律文书和框架以及相关的全球、区域和部门机构”。《联合国海洋法公约》（《公约》）和1995年《鱼类种群协定》等重要国际文书对捕鱼问题已经作出了详尽规定，现有的区域渔业管理组织或安排基本覆盖了所有公海海域。养护和可持续利用渔业资源应该继续由现有的渔业组织或安排来管理，其不应再作为新国际文书规范的事项。

关于海洋遗传资源获取，中方认为，原生境获取活动本质上属于《公约》规定的国家管辖范围以外区域的海洋科学研究，应适用自由获取制度，目的是激励海洋科研和创新。关于海洋遗传资源的惠益分享，中方建议采取务实的方法，将达成共识的内容先固定下来，以增强信心，继续扎实推进谈判工作。在充分照顾发展中国家关切和需求的基础上，优先考虑非货币化惠益分享机制。同时，中方对探讨建立货币化惠益分享机制持开放态度，也愿意参与讨论货币化惠益分享问题。

各方对海洋遗传资源的法律属性等问题还存在明显分歧。对于这些问题，中国代表团认为，各方应进一步加强磋商，不断寻求共识，努力实现养护和可持续利用国家管辖范围外区域海洋生物多样性的共同目标。在此，中方强调以下观点：

第一，关于定义。中国代表团认为，衍生物是生物化学合成的产物，不含有遗传功能单元。现有国际条约，如《生物多样性公约》《粮食和农业植物遗传资源国际条约》关于遗传资源的定义都没有包括衍生物，新国际文书有关海洋遗传资源的定义也不应包括衍生物。关于是否区分作为商品的鱼类和作为海洋遗传资源的鱼类。根据联大第69/292号决议，新国际文书“不应损害现有有关法律文书和框架以及相关的全球、区域和部门机构”。《公约》和1995年《鱼类种群协定》等国际条约对渔业问题已经作出详尽规定，现有的渔业管理组织和安排基本覆盖了所有公海海域。因此，养护和可持续利用渔业资源应该继续由现有的渔业组织或安排来管理，其不应再作为新国际文书规范的事项。

第二，关于海洋遗传资源的获取。中国代表团认为，原生境获取活动本质上属于《公约》规定的国家管辖范围外区域的海洋科学研究，应适用自由获取制度，以促进海洋遗传资源的开发和可持续利用。考虑到有关代表团提及就海洋遗传资源的获取建立追踪、许可和付费制度，中方认为，国家可在自愿基础上，将获取的海洋遗传资源信息通过信息交换的方式向国际社会适当通报。

第三，关于海洋遗传资源的惠益分享。海洋遗传资源惠益分享机制应鼓励海洋科学研究，促进全人类对海洋遗传资源的惠益分享。在充分照顾发展中国家关切和需求的前提下，应优先考虑非货币化惠益分享机制。同时，对探讨建立货币化惠益分享机制持开放态度。

二、关于划区管理工具

中国代表团支持由厄瓜多尔代表 77 国集团加中国所作的相关发言。中方看法如下：

划区管理工具包括多种管理形式和方法，不限于海洋保护区，新国际文书对其进行界定非常具有挑战性，中方建议总结现有划区管理工具的实践与经验，进行充分讨论。

如果定义，中方认为，该定义应至少包括三个要素：一是目标要素，划区管理工具的目标应与新国际文书的总体目标保持一致，即养护和可持续利用海洋生物多样性，这是划区管理工具的出发点和落脚点。在此基础上，具体运用包括海洋保护区在内的划区管理工具还需明确具体的保护对象和保护目标。根据一揽子协议和联大第 69/292 号决议的精神，划区管理工具与海洋遗传资源、环境影响评价、能力建设和技术转让等不可分割，对这些问题应平等对待、平行推进，孤立地强调其中某一个问题可能会背离本磋商的初衷。二是地理范围要素，划区管理工具应仅适用于国家管辖以外区域，相关保护区域应具有明确的地理界限。三是功能和手段要素。划区管理工具应包括不同功能和管理方法。

此外，关于海洋保护区设立的具体指导原则和方法，中方支持厄瓜多尔代表 77 国集团加中国所作有关发言，并进一步作以下几点说明：

第一，海洋保护区是工具而不是目标，设立海洋保护区要遵循必要性原则。

第二，设立海洋保护区要有完整的程序。刚才一些代表对有关程序提出了若干建议，包括申请、咨询、审查、决策、管理和监督等，这些都是很好的讨论基础。中方认为，设立海洋保护区提案应由国家提出。咨询的范围应尽可能广泛，所有利益攸关方都应参与其中。相关决策应坚持协商一致原则，只有在各方达成共识的前提下才建立相应保护区。

第三，关于新国际文书所规定的保护区与现有的全球性、区域性和部门性国际机构之间的关系，我们认为应遵循联大第69/292号决议的精神，即新国际文书不应损害现有有关法律文书和框架以及相关的全球、区域和部门机构。新国际文书应重在查漏补缺、填补空白，同时加强与现有划区管理工具的协调。

第四，设立海洋保护区应该有完整的保护措施和后续监督审查程序，不能只有建立，没有管理。

第五，设立海洋保护区应该明确的保护期限。这表面上看可能和某些国家的观点不太一致，但实质上并无不同。中方的主张是，经过评估和审查，如确有必要，期满的保护区可以延期。

关于划区管理工具的指导原则和方式，中方重申以下观点：

首先是区别保护原则。与新加坡代表提出的灵活性原则类似，中方认为应按照不同海域、生态系统、栖息地和种群的特点适用不同的工具予以保护。

其次是比例原则。技术层面上说，该原则要求保护措施与保护目标相适应，在符合成本效益原则的情况下应用划区管理工具。从更广泛的意义上说，该原则还要求考虑经济社会因素，马尔代夫、加拿大和菲律宾的代表都提到了这个问题。运用包括海洋保护区在内的划区管理工具需要专业和技术能力，为其制定一些标准固然很好，但应充分考虑发展中国家的关切和需求，不能对各国特别是发展中国家造成不当负担。

此外，针对划区管理工具，中方作了以下几点说明：

关于新国际文书设立海洋保护区的制度安排与现有国际文书和机制的关系。在非正式工作组讨论中，有关代表对于设立海洋保护区的制度安排提出了不同的建议和路径，包括建立全球性框架或者发挥区域的作用等。但建立海洋保护区的核心不是采取全球性框架还是区域性办法，关键是有关海洋生物多样性是否得到了应有的保护。我们认为，全球性框架与区域性办法的关系涉及新国际文书与区域性和部门性国际文书的关系问题，涉及不同的国家，应遵循条约关系的一般原则，予以慎重处理。

中国代表团认为，根据联大第 69/292 号决议规定，新国际文书应重在查漏补缺、填补空白，主要就 BBNJ 的养护和可持续利用问题作出规定，而不是另起炉灶。我们理解联大第 69/292 号决议包含两层含义，一是新国际文书应充分尊重现有的区域或部门机构的相关职能，不应干涉这些机构自主行使职能，包括自主决策权。二是对现有机构未能规管的海洋保护区，新国际文书可围绕养护和可持续利用 BBNJ 的总体目标作出规定。无论新国际文书就设立海洋保护区作出何种制度安排，都不能影响国际海底管理局、国际海事组织、区域渔业管理组织等现有机构的相关职能。但在具体设立海洋保护区过程中，新国际文书有必要与现有区域或部门机构加强合作与协调。

关于设立海洋保护区的条件和程序。中方认为，海洋保护区的设立应有明确的保护目标、确定的保护对象、具体的保护范围、适当的保护措施以及合理的保护期限等。同时，设立海洋保护区应遵循特定的程序，包括但不限于申请、咨询、审查、决策、管理和监督等。设立海洋保护区提案应由国家提出，一区一议、公开透明。咨询的范围应尽可能广泛，相关利益攸关方都可参与其中。最终决策应坚持协商一致原则，只有在各方达成共识的前提下才建立相应保护区。同时，保护区应该有一定的期限，如经过审查确有必要，期满的保护区可以延期。

三、关于环境影响评价

中国总体支持由厄瓜多尔代表 77 国集团加中国作出的有关发言。

关于环境影响评价的适用地理范围，中方同意阿尔及利亚代表非洲组发表的观点。根据《公约》第 194 条第 2 款“各国应确保……在其管辖或控制范围内的事件或活动所造成的污染不至大到其按照本公约行使主权权利的区域之外”，也就是说《公约》对国家管辖范围内的活动的跨界影响已作出规定。中方认为，在新国际文书中不必也不应针对就国家管辖范围内活动开展环境影响评价再另外作出规定。但是，对于发生在国家管辖范围外区域而可能对沿海国管辖海域产生影响的活动，应可以予以评估。

此外，关于环境影响评价的启动机制，中方认为，新国际文书作为《公约》的执行协定，其有关环境影响评价的制度安排应遵循《公约》所确立的基本法律框架和程序要素。《公约》第 206 条规定的环境影响评价的启动门槛标准应继续在新协定中予以适用。该标准就是“有合理依据认为”有关活动“可能造成重大污染或重大和有害的变化”。根据这个标准，对于一项具体的活动是否要开展环境影响评价，国家可按照《公约》规定并结合拟开展活动的具体情况，作出科学合理的判断。

关于是否对活动类型制定清单，中方认为，这种方式有一定局限性。一项活动对海洋环境的影响不仅取决于其类型，还取决于其规模、开展活动的位置及其对环境产生影响的方式。每项活动都有其特殊性，是否对有关活动开展环境影响评价，应该是一事一议。

关于新国际文书所规定的环境影响评价与现有机制的关系，

中方认为，现有国际文书已经就国家管辖范围以外区域中的某些活动作出了环境影响评价规定，如在公海深海捕鱼、国际海底区域深海矿产勘探、倾倒废物等领域，相关主管国际机构都有环境影响评价方面的要求和措施。特别是国际海底管理局在环境保护方面制定了详尽的规则和指南，进行环境影响评价等环保措施已是承包者在国际海底区域开展活动的必选动作。新国际文书应尊重这些机构或框架在本领域发挥的环评作用，避免就相同类型的活动设立新的环评标准和规则，造成不必要的重复或冲突。

关于开展环境影响评价的主体，根据《公约》第 206 条，评价主体应该是拟开展海洋活动的国家。国家按照一定的原则和标准，决定是否对其管辖或控制下的计划中的活动开展环境影响评价。环境影响评价的具体实施也应该由国家主导。环境影响评价所针对的活动是否可以继续进行，国家也应拥有最终决定权。同时，根据《公约》相关条款，国家应就环境影响评价的结果提交报告或予以公布，并应持续监测其所从事或准许的活动的影响，防止污染海洋环境。

关于环境影响评价的效率问题，中方认为，无论未来新国际文书对环境影响评价作何安排，相关程序设计应该便于操作，避免复杂烦琐，不能给各国特别是发展中国家施加不当负担。

针对战略环境影响评价，中方提出以下观点：

根据《公约》第 206 条，环境影响评价的对象是正在计划中的活动，“活动”一词的前面还有“计划中”这样一个限制词，这里所说的“活动”显然不包括“计划”，而是具体的活动。一些发达国家的实践显示，战略环境影响评价针对的是政策、计划和规划，这些都不是《公约》第 206 条所指的“活动”。

此外，《公约》第 206 条中还有一个单独的短语，即“在实际可行范围内”。战略环境影响评价的出发点可能是好的，它作为一种理念也许很有吸引力，一些发达国家亦不乏战略环境影响评价的实践。但是，对于包括中国在内的大多数发展中国家来说，开展战略环境影响评价以及与其密切相关的累积影响评价都不是一件易事。虽然环境保护是各国共同的义务，但也要顾及各国实际能力的差别。比如，在环境保护领域具有里程碑意义的《里约宣言》第 15 条便规定，各国应根据“自身能力”实施预防性原则。即使战略环境影响评价在国家层面有过成功实践，但各国的相关政策和规划千差万别，在国家管辖范围外区域实施战略环境影响评价是否“实际可行”，存在很大疑问。

未来随着科技的进步和国际社会整体能力的提升，在国家管辖范围外区域开展战略环境影响评价或许可以具备相应的条件。但至少在现阶段，中方认为，实施战略环境影响评价和累积影响评价的法律依据不足，可行性不强。

第一，关于环境影响评价的地理范围。大多数代表团认为，新国际文书关于环境影响评价的规定应仅针对发生在国家管辖范围以外区域的活动。中国代表团赞同这种观点，并认为《公约》第 194 条已对国家管辖范围内的活动的跨界影响作出规定，因此新国际文书不应就此作出规定。

第二，关于环境影响评价的启动标准。各方普遍认为《公约》第 206 条可以作为新国际文书关于环境影响评价规定的基础。根据该条，应由国家自行就拟开展的活动是否可能对海洋环境造成重大污染或重大和有害的变化作出判断。关于是否对活动类型制定清单，中方认为，每项活动对海洋环境的影响都不相同，采取清单列举的方式有一定的局限性。我们原则上不赞成采取清单方式。但如各方认为确有必要制定清单，我们认为该清单应是开放性的、建议性的，不具有法律拘束力，仅供各国参考。

第三，关于环境影响评价的主体。根据《公约》第 206 条，评价主体应该是拟开展海洋活动的国家。国家有权决定是否以及如何开展环境影响评价。同时国家应将环境影响评价的结果予以公布。

第四，关于战略环境影响评价。根据《公约》第 206 条，环境影响评价的对象是正在计划中的活动。战略环境影响评价针对的是政策、计划和规划，这些都不是《公约》第 206 条所指的“活动”。《公约》第 206 条还规定，环境影响评价应“在实际可行范围内”进行。在目前情况下，对于在国家管辖范围以外区域开展战略环境影响评价以及与其密切相关的累积影响评价是否“实际可行”尚存疑问。中国代表团认为，现阶段实施战略环境影响评价和累积影响评价的条件还不具备。

第五，关于新国际文书所规定的环境影响评价与现有机制的关系。中国代表团认为，现有国际文书已经就国家管辖范围以外区域中的某些活动的环境影响评价作出了规定，如在公海深海捕鱼、国际海底区域深海矿产勘探、倾倒废物等领域，相关主管国际机构都有环境影响评价方面的要求和措施。新国际文书应尊重这些机构或框架在本领域发挥的环评作用，避免就相同类型的活动设立新的环评标准和规则，造成不必要的重复或冲突。

四、关于能力建设和技术转让

中方支持厄瓜多尔代表 77 国集团加中国所作有关发言。能力建设和技术转让对养护和可持续利用国家管辖范围以外区域海洋生物多样性至关重要。能力建设和技术转让是 2011 年一揽子协议和联大第 69/292 号决议的主要问题之一。各方公认其同时也是一个跨领域问题。海洋遗传资源的获取和研究、海洋保护区的设立和管理以及开展环境影响评价等，都离不开经济社会支撑和专业技术能力。缺乏有效的能力建设和技术转让，有关活动可能还会只是少数国家的专利；新国际文书的制度架构将会失去根基；即使新国际文书能够最终出炉，其是否能被普遍接受也要打一个大大的问号。我们在《公约》能否被普遍接受方面不乏历史教训。1982 年《公约》生效后，很多发达国家因为不认同《公约》第十一部分的内容而拒绝批准《公约》，这种情况直到 1994 年相关执行协定出台后才有所改变。因此，就养护和可持续利用 BBNJ 的总体目标而言，无论怎么强调能力建设和技术转让的重要性都不为过。

能力建设和技术转让如此重要，它应该与其他议题一并统筹考虑。刚才，一些国家将能力建设和技术转让与海洋遗传资源的获取和惠益分享挂钩，欧盟也提到能力建设和技术转让的磋商取决于其他议题讨论的情况。考虑到能力建设和技术转让的重要性及其跨领域本质，中方同样认为有关该议题的讨论应与其他议题关联推进。如果缺乏有效的能力建设和技术转让，新国际文书对遗传基因资源、海洋保护区、环境影响评价等议题的实施和执行就会产生问题。一个国际文书如不具有执行性和操作性，再美好的制度和规则都无法发挥作用。

基于上述考虑，中方鼓励通过多种形式的国际合作来提升发展中国家养护和可持续利用国家管辖范围以外海洋生物多样性的能力和技术，充分利用包括政府间海洋学委员会等现有平台和机制的作用，同时也可考虑搭建新的信息共享平台。能力建设和技术转让的内容包括人才培养、信息交流、提升科研机构实力、促进科研基础设施建设、提供必要的技术装备和设施等方式。

无论是获取和研究海洋遗传资源，设立和管理海洋保护区，还是开展环境影响评价，都离不开经济社会支撑和专业技术能力。缺乏有效的能力建设和技术转让，不利于在全球范围内实现养护和可持续利用 BBNJ 的目标。在此，中国代表团观点如下：

第一，关于能力建设和技术转让的具体目标。新国际文书有关能力建设和技术转让的制度安排应是切实提升发展中国家，特别是小岛屿发展中国家、最不发达国家、内陆国和地理不利国以及有特殊利益需求的国家，在养护和可持续利用 BBNJ 方面的内生能力。为此目的，中国代表团支持能力建设须是“有意义的”，倡导既要“授人以鱼”，也要“授人以渔”。

第二，关于能力建设和海洋技术转让的方式。新国际文书应鼓励通过多种形式的国际合作开展相关能力建设和技术转让，包括搭建国际合作平台、双多边合作、发挥政府间海洋学委员会等现有相关国际组织的作用。

第三，关于能力建设和技术转让的内容。中国代表团认为，相关能力建设和技术转让应包括但不限于信息交流、人才培养、标准制定、提升科研机构实力、促进科研基础设施建设、提供必要的技术装备和设施等。

第四，关于信息交流所。中国代表团认为，该机制应公开透明，便于信息交流和共享。至于是利用现有信息交流机制，还是建立新机制，交流的具体信息有哪些，这些问题都应该在其他问题的讨论取得实质进展后进一步讨论确定。

五、关于跨领域问题

中国代表团支持厄瓜多尔代表 77 国集团和中国所作的发言，并作两点补充：

第一，关于新国际文书与《公约》的关系。中国代表团认为，新国际文书是在《公约》框架下制定的国际法律文件，应符合《公约》的目的和宗旨，应是对《公约》的补充和完善，不能偏离《公约》的原则和精神，不能损害《公约》建立的制度框架，不能损害《公约》的完整性和平衡性，应切实维护《公约》确立的国际海洋法律秩序，在《公约》所赋予的各项权利和义务之间保持平衡。正如中方在书面意见中提到的，中国代表团认为，1995 年《鱼类种群协定》第 4 条为确立新国际文书与《公约》的关系提供了指引。

第二，关于新国际文书的适用范围。联大第 69/292 号决议已经就此作出明确规定，即新国际文书应致力于“国家管辖范围以外区域海洋生物多样性的养护和可持续利用”，而且“不应损害现有有关法律文书和框架以及相关的全球、区域和部门机构”。这至少包括两方面意涵：在属地方面，新国际文书适用的区域是国家管辖范围以外的公海和国际海底区域，不及于国家管辖范围以内区域。在属事方面，新国际文书规范的活动应仅限于在上述区域的养护和可持续利用海洋生物多样性活动，而不是全部活动。

六、总结发言及关于下一步工作的建议

关于 BBNJ 养护和可持续利用问题，中国代表团重申了以下基本立场：

第一，中国代表团认为，预委会应严格遵守联大第 69/292 号决议的授权，“就根据《公约》的规定拟订一份具有法律约束力的国际文书的案文草案要点向大会提出实质性建议”。有关各方应严格遵循决议授权开展相关工作，预委会最终所提实质性建议应尽最大努力在协商一致的基础上反映各方共识。

第二，关于新国际文书与《公约》的关系，中国代表团认为，新国际文书是在《公约》框架下制定的国际法律文件，应符合《公约》的目的和宗旨，应是对《公约》的补充和完善，不能损害《公约》建立的制度框架，应在《公约》所赋予的各项权利和义务之间保持平衡。

第三，至于新国际文书与现行国际法和现有机制之间的关系，中国代表团认为，新国际文书不能与现行国际法以及现有的全球性、区域性和专门性的海洋机制相抵触，不能损害现有相关法律文书或框架以及相关全球性、区域性和专门性机构的职权，而应

促进与现有相关国际机构的协调与合作，避免职权重复或冲突。

第四，新国际文书的有关制度安排应有充分的法律依据和坚实的科学基础，并在 BBNJ 养护与可持续利用之间保持合理平衡。

第五，新国际文书应兼顾各方利益和关切，立足于国际社会整体和绝大多数国家的利益和需求，特别是应顾及广大发展中国家的利益，不能给各国尤其是发展中国家增加超出其承担能力的义务和责任。

第六，关于新国际文书跨领域问题的制度设计，不少代表团提出了诸多建议，包括缔约国会议、科学委员会等辅助机构、协定的监督和审查机制、争端解决机制等。中国代表团认为，对上述问题的讨论，应取决于各方对联大第 69/292 号决议一揽子问题讨论的结果和共识。有关制度安排总体上应具有可操作性、符合成本效益原则，避免对成员国，尤其是发展中国家造成过重负担。

根据联大第 69/292 号决议，预委会应在今年年底前向联大提交报告，就 BBNJ 养护和可持续利用国际协定的草案要素向联大提出实质建议。经过三次会议的讨论，各方均在一定程度上表达了各自的立场和关切，并在不少问题形成了一致理解。这些都为今年 7 月预委会第四次会议的成功举行奠定了良好基础。为切实落实和执行联大第 69/292 号决议，促进预委会完成其工作，中国代表团有如下建议：

第一，建议预委会主席在前三次会议特别是本次会议讨论的基础上，编纂一份包括各方观点和立场的非文件，在预委会第四次会议前向各方散发，以便下次预委会对相关问题的讨论更加深入和有效。

第二，建议预委会主席根据会议讨论情况就 BBNJ 国际协定草案的实质要素起草一份要素草案。要素草案的内容应全面客观反映各方讨论情况，包括各方取得共同理解的事项，以及分歧事项。该草案要素最好能在下次预委会举行前散发，以便各方提前研究。

Statement by the Chinese delegation on the third session of the Preparatory Committee on the negotiation of an international instrument on BBNJ

1. Marine Genetic Resources

China associates itself with the statement delivered by Ecuador on behalf of G77 and China. I would like to further explain China's views on the following issues.

China noted that in our discussions yesterday and today, many delegations referred to fishing, particularly whether to distinguish fish used as a commodity and fish valued for their genetic properties. According to the General Assembly Resolution 69/292, the new international instrument "should not undermine existing relevant legal instruments or frameworks as well as relevant global, regional and sectoral bodies." The United Nations Convention on the Law of the Sea and the 1995 United Nations Fish Stocks Agreement contain detailed provisions on fishing and existing Regional Fishery Management Organizations and Arrangements cover almost all of the areas of the high seas. The conservation and sustainable use of fishery resources should continue to be managed by these regional organizations or arrangements and should not be governed by the new international instrument.

On the issue of access to Marine Genetic Resources, China maintains that in essence, in-situ access falls within the scope of scientific research in areas beyond national jurisdiction as stipulated by the UNCLOS, therefore free access arrangements should be applied in order to encourage marine scientific research and innovation.

On the issue of benefit sharing of Marine Genetic Resources, we noted that both developed countries and developing countries support the discussion of non-monetary benefit sharing. In order to show practical results of our consultation, we suggest that a pragmatic approach could be taken to first solidify the points on which consensus has been reached so as to strengthen confidence and further advance this consultation process in a steady manner. We believe that priority consideration should be given to non-monetary benefit sharing mechanisms on the basis of full accommodation of the concerns and needs of developing countries. Meanwhile, we are open to exploring the possibility of establishing monetary benefit sharing mechanisms and are willing to take part in relevant discussion.

When the Informal Working Group on marine genetic resources (MGR) met, delegations held extensive and in-depth discussions on MGR, including its legal status, definitions, access and benefit sharing, and reached agreements on some issues. For example, all delegations expressed the view that the new international instrument should not hinder marine scientific research, endorsed the necessity of setting up a clearing house mechanism for the purpose of conservation and sustainable use of BBNJ and indicated their willingness to explore establishing a non-monetary benefit sharing mechanism. However, on some other issues, including the legal status of the MGR, there are still clear divergences. The Chinese delegation believes that all parties should further intensify consultation in search of an agreement so as to make the effort in achieving the common goal of conservation and sustainable use of marine biological diversity of areas beyond national jurisdiction.

To this end, the Chinese delegation wishes to emphasize the following points:

First, on definition. The Chinese delegation is of the view that derivatives, as the product of a biochemical synthesis process, are without functional units of heredity. The definition of genetic resources in the existing international treaties, such as the Convention of Biological Diversity (CBD) and the International Treaty on Plant Genetic Resources for Food and Agriculture (ITPGRFA), makes no mention of derivatives. The new international instrument should not include derivatives in the definition of marine genetic resources, either. As for whether to make a distinction between fish used as commodity and fish used as marine genetic resources, the new international instrument "should not undermine existing relevant legal instruments and frameworks and relevant global, regional and sectoral bodies" according to GA Resolution 69/292. International treaties, such as the United Nations Convention on the Law of the Sea (UNCLOS) and the 1995 Fish Stocks Agreement, have detailed provisions on issues relating to fisheries. The existing regional fisheries management organizations or arrangements (RFMO/As) basically cover all areas of the high seas. Therefore, the conservation and sustainable use of fishery resources should continue to be managed by the existing RFMO/As and should not become an element to be regulated by the new international instrument.

Second, on access to marine genetic resources. The Chinese delegation is of the view that in-situ access in essence belongs to the marine scientific research conducted in areas beyond national jurisdiction, as stipulated by the Convention. As such, the free-access regime should apply, which will contribute to the exploitation and sustainable use of marine genetic resources. Given that some delegations have put forward proposals on setting up a system that entails traceability, permits and payments for accessing MGR, my delegation is of the opinion that States can, on a voluntary basis, provide due notification on accessed marine genetic resources through information exchange.

Third, on sharing of benefits from the marine genetic resources. The mechanism for such benefit sharing should encourage marine scientific research and facilitate the sharing of benefits from marine genetic resources for all mankind. The Chinese delegation is of the view that on the premise of fully accommodating the concerns and needs of developing States, the PrepCom should give priority to non-monetary benefit sharing mechanism. Meanwhile, we are open to discussing and exploring the establishment of a monetary benefit sharing mechanism.

2. Area-based management tools

The Chinese delegation associates itself with the statement made by Ecuador on behalf of the Group of 77 and China. Here I wish to further elaborate on China's views.

Area-based management tools include various management forms and approaches besides marine protected areas. It is very challenging to arrive at a definition in the new international instrument. China suggests that we look at the practices and experiences of existing ABMTs and conduct thorough discussions on this issue.

If there is to be a definition, China believes that it should at least include the following three elements:

(1) The element of objective. The objective of area-based management tools should be consistent with the goal of the new international instrument, namely, conservation and sustainable use of marine biodiversity. This should be the whole purpose of ABMTs. On that basis, when applying specific area-based management tools, including marine protected areas, it is necessary to identify the objects and targets of protection. Here I wish to stress that in the spirit of the 2011 package agreements and the GA resolution 69/292, issues of ABMTs, marine genetic resources, environmental impact assessment, capacity building and transfer of technology are inseparable. Therefore it is necessary to take a holistic approach in discussing these issues and push them forward in parallel. Singling any issue out to isolate it from the others may run counter to the intention of this consultation.

(2) The element of geographical scope. The area-based management tools should be applied only in areas beyond national jurisdiction and the relevant protected areas need to have clear geographical delineation.

(3) The element of function and means. The area-based management tools should include different functions and means of management.

Regarding the guiding principles and approaches for the creation of MPAs, China supports what Ecuador stated on this topic on behalf of Group 77 and China. In addition, we would like to make the following points:

First, MPAs are a means to an end; they are not the end. The establishment of MPAs should follow the principle of necessity.

Second, a complete set of procedures must be in place for the creation of MPAs. We have heard a range of proposals from some delegations on such procedures, covering submission of proposals, consultation, review, decision-making, management and monitoring. We are open to continued discussion on this basis. In our view, it makes more sense for states to be the submitters of proposals on the creation of MPAs. The scope of consultations should be as broad as possible, to allow the participation of all stakeholders. The consensus principle should apply to decision-making, that is, protected areas are created only on the basis of consensus among the parties concerned.

Third, as regards the relationship between the MPAs regulated by the new international instrument and the existing global, regional and sectoral bodies, it is our view that we should be guided by GA resolution 69/292, which states "the new instrument may not undermine existing relevant legal instruments and frameworks and relevant global, regional and sectoral bodies". The priority of the new international instrument is to identify and fill the gaps and shortfalls, while strengthening the coordination with existing ABMTs.

Fourth, the creation of MPAs should be supported by a complete set of protection measures and follow-up monitoring and review procedures. Creation without management is not the way to go.

Fifth, China believes time limits for protection should be placed on protected areas. At first blush, this idea may appear to be at variance with the views of some delegations, but in actual fact, it is not. If evaluation and review find continued protection necessary for a MPA about to expire, an extension may be granted to this area and its protection measures.

The Chinese delegation has spoken on all the three items you are in charge of, which is ample proof of our support to your work.

In its written comments submitted previously, China elaborated its views on guiding principles and approaches of ABMTs. Here, I believe there is a need to reiterate some of our views as our contribution to the discussion and part of our effort to expand consensus.

First, on the principle of necessity. In line with the rationale of the principle of flexibility proposed by the delegate of Singapore, China believes that tools of protection need to be tailored to the characteristics of different sea areas, eco-systems, habitats and species.

Second, on the principle of proportionality. At the technical level, the principle of proportionality means that protective measures need to be suitable to the objects of protection and that ABMTs should be applied in a cost-effective way. In a broader sense, it means that economic and social factors need to be taken into account. We support the views of the Maldives, Canada and the Philippines on this issue.

Application of ABMTs, including MPAs, also requires a certain level of professional capacity. While it is desirable to develop standards for ABMTs, including MPAs, it is also necessary to take into full consideration the concerns and needs of developing countries and avoid creating excessive burden for countries, especially developing countries.

The Chinese delegation would like to thank Madam Facilitator Alice Revell for her professional, lucid report. Our thanks also go to our chair, Ambassador Duarte, for his clear guidance to structure our discussion. The Chinese delegation would like to make the following points:

On the relationship between the institutional arrangement for the designation of MPAs under the new international instrument and the existing international instruments and mechanisms. During our discussion at the informal working group, some delegations came up with various proposals and approaches regarding the institutional arrangement for the designation of MPAs. They included, among others, creating a global framework or giving regions a role to play. But the question central to the designation of MPAs is not whether we opt for a global framework or a regional approach. The key question here is whether or not the relevant marine biodiversity is accorded due protection. In our view, the relationship between a global framework and a regional approach is linked to the relationship between the new international instrument and the region- and sector-specific international instruments. Different states are involved. The general principles governing treaty relations shall apply. Caution is advised.

My delegation is of the view that, based on the GA resolution 69/292, the new international instrument should focus on identifying and making up for gaps and shortfalls, by articulating provisions on the conservation and sustainable use of BBNJ, rather than creating a separate system from scratch. To our mind, the thrust of resolution 69/292 is twofold: On one hand, the new international instrument should fully respect the relevant mandates of the existing regional or sectoral bodies and should not interfere with the autonomous functioning of these bodies, including their independent decision-making powers. On the other hand, for MPAs not covered by the existing bodies, the new international instrument may introduce provisions centering on the overarching goal, i. e. the conservation and sustainable use of BBNJ. Whatever institutional arrangement we end up with for the designation of MPAs under the new international instrument, it may not prejudice the relevant mandates of existing bodies, like ISA, IMO and RFMOs. But as regards the actual process of designating MPAs, strengthened cooperation and coordination between the new international instrument and the existing regional or sectoral bodies is required.

On the conditions for, and procedures of, the designation of

MPAs. To designate an MPA, there should be an explicit objective of protection, a well-defined target of protection, a specific scope of protection, an appropriate set of protection measures and a reasonable term of protection. Moreover, the designation of an MPA should follow dedicated procedures, including but not limited to submission, consultation, review, decision - making, management and monitoring. The Chinese delegation is of the view that any proposal for the designation of an MPA should be submitted by the State and subsequently considered on a case-by-case basis in an open and transparent manner. Consultations should be as broad as possible and open to all relevant stakeholders. The principle of consensus should prevail when final decisions are made, which means no MPA may be created until all parties concerned have come to a consensus. Furthermore, a time limit should be placed on each MPA, renewable upon expiration following a review that finds such renewal sufficiently necessary.

3. EIA

On the whole, China aligns itself with the relevant statement made by Ecuador on G77 and China.

On the geographical scope of environmental impact assessments, China agrees with the view expressed by Algeria on behalf of the African Group. Paragraph 2 of Article 194 of the UNCLOS stipulates that "*States shall … ensure that … pollution arising from incidents or activities under their jurisdiction or control does not spread beyond the areas where they exercise sovereign rights in accordance with this Convention.*" So the Convention already contains provisions on the transboundary impact of activities in areas under national sovereignty. That being so, it is the view of

China that there is neither need nor reason for the new international instrument to make further provisions on environmental impact assessments of activities in areas under national sovereignty. However, assessments can be made of activities beyond national jurisdiction that may produce impact on areas under national jurisdiction of coastal states.

Regarding the trigger for environmental impact assessment, China is of the view that the new international instrument, being an implementing agreement of the UNCLOS, should have an institutional arrangement for EIA that stays in line with the basic legal framework and procedural elements established under the UNCLOS. Article 206 of the Convention provides that EIA be triggered only when there is "reasonable ground for believing" that such activities "may cause substantial pollution of or significant and harmful changes to the marine environment". This provision should also apply to the new international instrument, under which the decision to perform EIA or otherwise is for states to make. States are to make a scientific judgement on the environmental impact of a planned activity pursuant to the provisions of the Convention and in light of the circumstances of this activity.

Regarding the listing of activities to which EIA should apply, China thinks this approach has some limitations. The impact of a given activity on the marine environment is a function not only of its type, but also of its size, the location of its operation and the way in which it impacts the environment. Every activity has its unique characteristics. It is advisable to take a case-by-case approach when deciding whether or not EIA should be performed for a given activity.

As regards the relationship between the EIA provisions in the new agreement and the existing mechanisms, the existing international instruments already contain EIA provisions on ABNJ activities. For example, the competent bodies have EIA measures in place for, *inter alia*, high sea fishing, deep-sea mineral exploration and waste dumping. In particular, the International Seabed Authority has detailed EIA rules and guidelines for the exploitation of resources in the Area, to the extent that EIA is an indispensable part of any activity undertaken by the contractors. We should encourage these bodies to keep up the role they play in EIA and should avoid any overlap or conflict between the new international instrument and the existing regulations.

Just now, many delegations presented highly thought-provoking proposals on EIA processes. On this topic, China would like to highlight two further points:

The first is about the actor that performs EIA. Under Article 206 of the UNCLOS, the subject of EIA should be the state that plans to conduct marine activities. The state, on the basis of certain principles and criteria, decides whether or not to conduct EIA on a planned activity under its jurisdiction or control. The state should also have the authority to decide whether the activity on which EIA has been performed may continue. As required by the UNCLOS, the state shall additionally prepare a report on the findings of the EIA and make it available, and continue to monitor the impact of this activity on the environment, to prevent any substantial pollution of or significant and harmful changes to the marine environment.

The second is about the efficiency of EIA. China shares the Russian delegate's concern on this issue. Irrespective of the type of arrangement the new international instrument will have for EIA, we believe the relevant procedures should be so designed as to be operation-friendly without being overly elaborate and complex, to avoid overburdening the states in general and developing countries in particular.

China attaches great importance to capacity building and technology transfer in relation to environmental impact assessments. Conscious of the time, and in view of the cross-cutting nature of this issue, we would like to confine this intervention to strategic environmental assessments (SEAs).

In the earlier discussion, a good number of delegates referred to Article 206 of UNCLOS. We have noted that by and large, no objection to this article has been voiced by any country. Rather, the majority view is that this article should form the basis and starting point of the institutional arrangement on EIA under the new international instrument. According to Article 206, the subject matter of EIA is planned activities-the word "activities" is preceded and modified by the adjective "planned". It is self-evident that the "activities" referred to therein are particular activities that do not cover "planning". The practice of some developed states indicates that SEAs relate to policies, plans and programs, none of which fall within the purview of "activities" under Article 206.

Furthermore, China would like to call your attention to a parenthetical phrase in Article 206, i. e. "as far as practicable". The intention behind SEAs may be a positive one and as a concept, it may be rather appealing. There is no shortage of practice in SEAs among some developed states. Nevertheless, for the majority of developing states, China included, conducting SEAs and closely related cumulative impact assessments is no easy task. Although environmental protection is a common obligation on all States, the disparities among the States in terms of actual capacities should be taken into account. For example, Principle 15 of the*Rio Declaration*, a milestone instrument on environmental protection, provides that "the precautionary approach shall be widely applied by States according to their capabilities". Notwithstanding the fact that SEAs may have succeeded at the national level, the relevant policies and programs of different States are poles apart. It is very questionable whether SEAs in ABNJ are "practicable".

Future progress in science and technology and improvement of the international community's capacity as a whole may create the necessary conditions for SEAs in ABNJ. But it is China's view that, for now at least, SEAs and cumulative impact assessments lack sufficient legal ground and are not very feasible.

The Chinese delegation would like to make the following elaborations:

First, on the geographical scope of environmental impact assessments. The Chinese delegation has noticed that most delegations are of the view that provisions on environmental impact assessments in the new international instrument should only govern activities in areas beyond national jurisdiction. My delegation agrees with this view and believes that since Article 194 of the UNCLOS already stipulates on the trans-boundary impact of activities within areas of national jurisdiction, there is no need for the new international instrument to make provisions in that regard.

Second, on the criteria to trigger environmental impact assessments. It is the general view that Article 206 of the UNCLOS can be the basis of provisions on environmental impact assessment in the new international instrument. In accordance with that Article, it is up to states themselves to determine whether the planned activities may cause substantial pollution of or significant and harmful changes to the marine environment. As for whether there should be a list of activities, the Chinese delegation is of the view that since activities vary in their impacts on the marine environment, there are certain limitations to the listing approach. In principle, we are not in favor of listing, but if parties deem it absolutely necessary to have a list, we expect it to be open, indicative and non-legally binding, serving only as reference to states.

Third, on who should conduct environmental impact assessments. According to Article 206 of the UNCLOS, states planning to undertake marine activities should be responsible for the assessments. States have the right to decide whether and how to conduct environmental impact assessments. They should also make publicly available results of those assessments.

Fourth, on strategic environmental impact assessments. As provided in Article 206 of the UNCLOS, the object of environmental impact assessments is the planned activities. Strategic environmental impact assessments target policies, plans and programs which do not constitute "activities" referred to by Article 206. Article 206 also stipulates that environmental impact assessments should be made "as far as practicable". Under the current circumstances, it remains questionable whether strategic environmental impact assessments for areas beyond national jurisdiction and the closely-related assessments of cumulative impacts are practicable or not. The Chinese delegation is of the view that at this stage, conditions are not yet in place to allow for the implementation of strategic environmental impact assessments and cumulative impact assessments.

Fifth, on the relations between the provisions on environmental impact assessment in the new international instrument and existing mechanisms. The Chinese delegation observes that existing international instruments contain regulations on environmental impact assessments of some activities in areas beyond national jurisdiction. For example, in the areas of deep sea fishing in the high seas, deep sea mineral exploration in the international seabed area and waste dumping, all relevant competent international organizations have requirements and measures for environmental impact assessments. The new international instrument should respect the role of these institutions or frameworks in relation to environmental impact assessment in their respective domains and refrain from making new standards and rules of environmental impact assessment for the same types of activities, so as to avoid unnecessary duplications or conflicts.

4. Capacity Building and Transfer of Marine Technology (CB/TMT)

China endorses the statement delivered by Ecuador on behalf of G77 and China. Capacity-building and technology transfer is of vital importance to the conservation and sustainable use of marine biodiversity beyond areas of national jurisdiction. It is one of the issues addressed in the 2011 package agreement and GA Resolution 69/292. It is also recognized by all parties as a cross-cutting issue. The acquisition and study of marine genetic resources, the establishment and management of MPAs and the conduct of environmental impact assessments cannot be accomplished without economic and social backing or expertise and technical capacity. Absent effective capacity-building and technology transfer, the related activities would remain the proprietary domain of a small number of countries, the institutional framework of the new international instrument would be without foundation and, even if the instrument eventually were to come to light, its universal acceptance would also be subject to a big question mark. As for the ability or otherwise to gain universal acceptance, there is no lack of negative lessons in the past. After the 1982 UN Convention on the Law of the Sea came into force, many developed countries refused to ratify the Convention as they did not identify with what was contained in Part XI of the Convention. This situation remained unchanged until the 1994 Implement Agreement was adopted. In view of this, for the overarching goal of the conservation and sustainable use of BBNJ, the importance of capacity-building and technology transfer cannot be over-emphasized.

Such is the importance of capacity-building and technology transfer that it should be considered in conjunction with other agenda items. Just now, some countries linked capacity-building and technology transfer with the acquisition of marine genetic resources and benefit sharing. The European Union also said that the consultations on capacity-building and technology transfer hinge on discussions on other agenda items. Given the importance and cross-cutting nature of capacity-building and technology transfer, China also believes that discussions on this item should be advanced along with those on other agenda items. Without effective capacity-building and technology transfer, the implementation and execution of the new international instrument in such areas as genetic resources, marine protected areas and environmental impact assessment would be problematic. If an international instrument is not operable or executable, regimes and rules, however perfect they may be, cannot possibly have any effect.

Based on the above considerations, China encourages multiple forms of international cooperation to improve developing countries' capacity and technology for the conservation and sustainable use of marine biodiversitybeyond areas of national jurisdiction. Full use should be made of the role of existing platforms and mechanisms, including the Intergovernmental Oceanographic Committee. Consideration may also be given to building a new clearinghouse platform. Capacity-building and technology transfer include the training of personnel, exchange of information, elevation of the institutional capacity in scientific research, promotion of the development of scientific research infrastructure, and provision of the requisite technical equipment and facilities, among others.

The Chinese delegation wishes to thank Mme. Rena Lee for her hard work in facilitating the Informal Working Group on Capacity Building and Transfer of Marine Technology (CB/TMT) and Mr. Chair for the guidance provided on this agenda item.

As emerged in the Informal Working Group discussion, there is the general recognition that CB/TMT is a cross-cutting issue. Whether it is the access and study of MGRs, or the establishment and management of MPAs, or the undertaking of EIAs - non of these can be accomplished without the socioeconomic support as well as professional and technical know-how. The absence of effective CB/TMT would not serve the interest of the global goal of conservation and sustainable use of BBNJ. In this connection, my delegation would like to emphasize the following points.

1. *On the specific objectives of CB/TMT.* The institutional arrangement under the new international instrument should be set up with the view of effectively upgrading the native capacity on the part of Developing Countries, in particular small island developing countries, least developed countries, landlocked developing countries and geographically disadvantaged countries as well as countries with special interest and needs, to conserve and sustainably use BBNJ. To this end, my delegation endorses the idea of "meaningful" capacity building, and advocates "giving people fish" and, more importantly, "teaching people how to fish".

2. *On the modalities of CB/TMT.* The new international instrument should encourage various forms of international cooperation for CB/TMT, including establishing international platforms for collaboration, conducting bilateral and multilateral co-operations and giving existing international organizations, such as the Intergovernmental Oceanographic Commission of UNESCO, a role to play.

3. *On the elements of CB/TMT.* The Chinese delegation is of the view that CB/TMT should include, but not limited to information exchange, human resources training and development, standards setting, elevation of institutional R&D capacity, R&D infrastructure, provision of the requisite technical equipment and facility, etc.

4. *On clearing-house mechanism (CHM)*. As can be observed in the discussion, most countries see the need to have a provision in the new international instrument to regulate the CHM, which, in the view of my delegation, should be open and transparent, so as to facilitate the sharing and exchange of information. As to whether to utilize the existing CHMs or create a new CHM as well as what specific information is to be included in the exchange, such questions can be determined upon further discussion once substantive progress has been achieved on other issues. The Chinese delegation stands ready to join others in continuing the in-depth discussion on this topic.

5. Areas Beyond National Jurisdiction

My delegation aligns itself with the statement delivered by Ecuador on behalf of G77 and China and would like to make the following two additional comments.

First, on the relationship between the new international instrument and the Convention. The Chinese delegation is of the view that the new international instrument is an international legal document developed under the framework of the Convention. It must conform to the purposes and objectives of the Convention and further supplement and improve the Convention. It cannot deviate from the principles and the spirit of UNCLOS, erode the institutional framework thereof and undermine the integrity and balance of the Convention. Rather, it should effective uphold the international maritime legal order and maintain the balance between the rights and obligations conferred by the Convention. As stated in the written submission, in the view of the Chinese delegation, Article 4 of the 1995 UN Fish Stocks Agreement provides some guidance on addressing the relationship between the new instrument and the Convention.

Second, on the applicable scope of the new international instrument. The Chinese delegation is of the view that GA Resolution 69/292 has already made it very clear that the new international instrument should focus on "the conservation and sustainable use of biological diversity of areas beyond national jurisdiction" and it "should not undermine existing relevant legal instruments and frameworks and relevant global, regional and sectoral bodies". This implies that (1) in terms of geographical scope, the new international instrument applies to high seas and international seabed areas beyond national jurisdiction, excluding areas within national jurisdiction; and (2) in terms of material scope, the activities regulated by the new intentional instrument are only those pertaining to the conservation and sustainable use of marine biological diversity in the areas mentioned above. In other words, not all activities are covered.

6. The work of the Preparatory Committee in the next phase

While associating itself with the statement by the distinguished representative of Ecuador on behalf of the Group of 77 and China, the Chinese delegation would like to take this opportunity to reiterate our basic positions on the conservation and sustainable use of BBNJ and to share our thoughts on the way forward:

One, regarding the conservation and sustainable use of BBNJ, my delegation would like to reiterate our basic position as follows:

First, in my delegation's view, the PrepCom should abide by, to the letter, the mandate contained in GA resolution 69/292, namely, "to make substantive recommendations to the General Assembly on the elements of a draft text of an international legally binding instrument under the Convention". All parties concerned should conduct relevant work in strict accordance with this mandate and the substantive recommendations from the PrepCom as its final outcome should, to the greatest extent possible, reflect the points of convergence achieved by consensus.

Second, regarding the relationship between the new international instrument and the Convention, my delegation is of the view that the new international instrument is an international legal instrument developed within the framework of the Convention. As such, it should stay in line with the purposes and objectives of the Convention, supplementing and complementing the latter. It may not undermine the institutional framework created by the Convention. A balance should be maintained between various rights and obligations enshrined in the Convention.

Third, regarding the relationship between the new international instrument and the existing international laws and existing mechanisms, my delegation is of the view that the new international instrument may not contravene or contradict the existing international laws or existing global, regional and sectoral seas mechanisms, nor can it undermine the relevant existing legal instruments or frameworks or the mandates of relevant global, regional and sectoral bodies. Rather, it should facilitate coordination and cooperation with the relevant existing international bodies while avoiding overlap or conflict of mandates.

Fourth, the relevant institutional arrangements under the new international instrument should be grounded in sound legal justifications, solid scientific evidence and a proper balance between BBNJ conservation and its sustainable use.

Fifth, the new international instrument should accommodate the whole spectrum of interests and concerns of various parties and base itself on the interests and needs of the international community as a whole and of the vast majority of states. In particular, it should accommodate the interests of the wide developing world and may not impose such obligations and responsibilities on the states, especially developing states, that exceed their capacity to cope.

Sixth, regarding the institutional design for the cross-cutting issues under the new international instrument, a good number of delegations came up with a host of proposals, concerning COP, a scientific committee and other subsidiary bodies; the monitoring and review mechanism for the agreement; the dispute settlement mechanism; and so on. My delegation is of the view that discussion of those matters hinges on the outcome and consensus emanating from the discussion of the whole package under resolution 69/292. Any institutional arrangement should, as a whole, be actionable and cost-effective, without overburdening member states, especially developing states.

Two, the way forward.

According to GA resolution 69/292, the PrepCom is mandated to submit to the General Assembly by the end of this year a report that contains substantive recommendations on the elements of a draft text of an international legally binding instrument under the Convention. Discussions at the three sessions have enabled all parties to articulate, to some extent, their positions and concerns and developed a consensus on a wide range of issues. This paves the way for the success of the upcoming fourth session of the PrepCom in July this year. To operationalize and implement GA resolution 69/292 and help the PrepCom complete its work, my delegation would like to propose the following:

First, we encourage the Chair of the PrepCom to prepare a non-paper that reflects the views and positions of all parties, on the basis of the discussions at the first three sessions, especially at this session, to be distributed to all parties prior to the fourth session, to enable more in-depth and effective discussions on the relevant matters at the next session.

Second, we encourage the Chair of the PrepCom to prepare a draft text of substantive elements for an international agreement on BBNJ, again based on the discussions at the successive sessions. This elements draft should reflect the discussions in a comprehensive and objective manner, including points of convergence and points of divergence. Preferably this draft is made available for distribution prior to the next session of the PrepCom, so that delegations can study it in advance.

BBNJ 国际协定谈判预备委员会第四次会议中国代表团发言汇编

国家管辖范围以外区域海洋生物多样性（BBNJ）养护和可持续利用问题国际协定谈判预备委员会第四次会议于 2017 年 3 月 27 日在纽约联合国总部举行。此次会议系预委会阶段最后一次会议，各方高度重视。会议继续就新国际文书所涉海洋遗传资源及其惠益分享、海洋保护区等划区管理工具、环境影响评价、能力建设和技术转让以及跨领域问题等进行了深入讨论，并最终通过了包含新国际文书草案要素的预委会工作报告。会上，中国代表团建设性参与讨论，发言有理有据，在多个关键问题上发挥定调和引领作用，不仅有效维护我国家利益，也获得不少与会代表的称赞。现将代表团团长、外交部条法司马新民副司长在预委会第四次会议上的主要发言内容摘录如下，以备工作参考。

一、一般性发言

中国代表团支持厄瓜多尔代表 77 国集团加中国所作的发言。中国代表团感谢主席先生和秘书处同事所提出的《主席指引建议》。该建议纳入了各方在预委会前三次会议和书面意见中所表达的观点，内容总体平和，较为客观地反映了此前的磋商成果，为本次会议的讨论提供了良好基础。中国代表团愿与各方一道，一如既往地支持主席先生的工作，以建设性的态度积极参与讨论。

中国政府高度重视BBNJ的养护和可持续利用问题，并于今年4月20日提交了修订版书面意见。中国代表团愿在此重申，按照联大第69/292号决议，预委会的职权是“就根据《联合国海洋法公约》（《公约》）的规定拟订一份具有法律约束力的国际文书的案文草案要点向大会提出实质性建议”，有关各方应严格按照联大决议授权开展相关工作。中国代表团相信，在主席先生的引领下，预委会定能严格依照联大决议的授权和要求，按时完成本阶段的工作任务。

二、海洋遗传资源及其惠益分享

中国代表团赞同厄瓜多尔代表77国集团加中国所表达的立场，支持各方就海洋遗传资源的获取和惠益分享问题进一步加强磋商，不断寻求共识，努力实现养护和可持续利用BBNJ的共同目标。对此，中国代表团愿强调以下几点：

第一，关于惠益分享的目标。中国代表团建议增加两点内容：一是“有利于维护国际社会的共同利益，促进人类共同福祉”，二是“有利于兼顾养护和可持续利用国家管辖范围以外区域海洋遗传资源”。

第二，关于知识产权问题。根据联大第69/292号决议，BBNJ新国际文书不应损害现有有关法律文书和框架以及相应的全球、区域和部门机构。由于知识产权问题，特别是遗传资源的来源披露等问题，正由世界知识产权组织和世界贸易组织等讨论，中国代表团认为，新国际文书没有必要对此作出规定。

第三，关于监测海洋遗传资源的利用问题。考虑到各方对海洋遗传资源利用概念的范围尚未达成共识，中国代表团建议删去相关内容。

三、海洋保护区等划区管理工具

中国代表团总体支持厄瓜多尔代表 77 国集团加中国所作相关发言，并愿就该问题进一步作以下几点评论和说明：

第一，关于一体化管理办法。中国代表团愿作以下解释。按照《公约》序言要求，“意识到各海洋区域的种种问题都是彼此密切相关的，有必要作为一个整体来加以考虑”，这是“一体化管理办法”的法律依据。所谓“一体化管理办法”，即将海洋区域作为一个整体来考虑，决定是否及如何适用适当的划区管理工具，以弥补目前按海域和国际组织职权来设置保护区的做法。

第二，关于区域的识别。中国代表团认为，识别划区管理工具适用的区域应综合考虑科学依据、社会经济因素和不同区域的具体情况等，加以确定。所谓科学依据，就是划区管理工具的认定应有客观科学标准，基于最佳科学证据，并应遵循公开透明等科学程序。所谓社会经济因素，就是划区管理工具的运用需考虑海洋遗传资源可持续利用的目标，顾及各国的社会和经济发展需要。所谓不同区域的具体情况，就是运用划区管理工具要根据不同海域、生态、栖息地和种群的特点，具体问题具体分析，逐案处理。中国代表团建议，宜参照现有国际文书有关确定划区管理工具标准的做法，在新国际文书中对此不作具体规定，而由日后新国际文书框架下设立的专门机构来制定相关标准。

第三，关于划区管理工具的时限。中国代表团认为，划区管理工具包括海洋保护区旨在实现特定保护目标，一旦保护目标实现，划区管理工具包括海洋保护区就应终止。中国代表团建议，应对设立划区管理工具包括海洋保护区的时限作出规定。在保护时限届满后，视评估情况，决定是终止还是延续划区管理工具包括海洋保护区。

四、环境影响评价

中国代表团支持厄瓜多尔代表 77 国集团加中国所作发言，并重申新国际文书的有关规定应与《公约》第 206 条等规定严格保持一致。在此，中国代表团愿作以下几点评论：

第一，关于所谓“邻近原则”。中国代表团注意到，一些邻近国家管辖范围以外区域的沿海国要求规定所谓“邻近原则”。我们完全理解有关国家的关切，不反对在具体规定中有所反映，如设立海洋保护区和进行海洋环境影响评价时考虑或咨询邻近沿海国的意见，但不认为国际法中存在所谓“邻近原则”，也不支持将其作为新国际文书的一般原则。中国代表团认为，“邻近原则”没有法律依据。根据《公约》，各国在国家管辖范围以外区域具有同等的权利和义务，任何国家包括沿海国都不具有优先的权利。如果新国际文书规定所谓“邻近原则”，将打破《公约》关于沿海国和其他国家之间的权利义务平衡，侵蚀公海自由和国际社会的整体利益。我们认为，《公约》已确定“适当顾及”原则，按照《公约》精神，国家在公海和国际海底区域开展活动时，应“适当顾及”其他国家包括邻近沿海国的权利和自由。因此，“适当顾及”原则已可满足有关邻近沿海国的关切，希望有关国家考虑适用“适当顾及”原则主张有关权利。

第二，关于战略环评。战略环境评价针对的是政策、计划和规划等，这些都不是《公约》第 206 条所指的“活动”。从现实情况来看，各国对战略环评认识尚不一致，也缺乏相对成熟的实践，在国家管辖范围以外区域开展战略环评是否是“实际可行”的，存在很大疑问。因此，中国代表团认为，新国际文书不宜就战略环评作出规定。

第三，关于累积环评。中国代表团认为，根据各国实践，“累积环评”并非单一的环评类型，可考虑将累积影响作为对活动的环境影响评价的考虑因素之一。

第四，关于监测和审议。当前案文对环评监测和审议的主体语焉不详，各方对此存在不同意见。中国代表团建议在实质性要素建议中删去有关内容。

五、跨领域问题

中国代表团总体支持厄瓜多尔代表77国集团和中国所作发言，并愿就以下问题提出建议并作出评论：

第一，关于序言要素。中国代表团建议在序言要素中增加两项内容。一是“承认人类在养护和可持续利用国家管辖范围以外的海洋生物多样性方面已成为一个不可分割的命运共同体”。二是“承认国际社会整体在养护和可持续利用国家管辖范围以外区域海洋生物多样性方面具有共同利益”。

第二，关于指导原则和方法。善意原则是国际法上公认的一般法律原则。同时，《公约》第300条也规定：“缔约国应诚意履行根据本公约承担的义务并应以不致构成滥用权利的方式，行使本公约所承认的权利、管辖权和自由。”新国际文书作为《公约》框架下的新执行协定，显然也应适用善意原则。中国代表团建议在指导原则和方法中加入“善意原则”。

第三，关于信息交换所。中国代表团支持建立统一、高效、便捷的信息交换所制度，以方便用户以安全、灵活和务实的方式交换信息。同时，中国代表团认为知识产权、商业秘密等信息不应在信息交换范围之内。

第四，关于责任和赔偿责任。国际法委员会是联合国大会下设的一个专家团体，其制定的《国际不法行为国家责任条款草案》

未经国家审议通过，虽然某些条款反映了习惯国际法，可作为习惯国际法存在的证据，但条款草案本身并不构成习惯国际法。国际实践中，鲜见国际条约直接援引国际法委员会条款草案作为法律依据的做法。因此，中国代表团认为新国际文书草案要素不应援引国际法委员会《国际不法行为国家责任条款草案》。

六、下步工作

中国代表团支持厄瓜多尔代表 77 国集团加中国的发言，认为预委会应严格按照联大第 69/292 号决议要求开展工作，并愿借此机会就下步工作提出以下几点建议：

第一，关于预委会向联大提交的报告。按照联大第 69/292 号决议要求，预委会应在今年年底前向联大报告其工作进展情况。中国代表团认为，有关报告应全面、客观反映预委会讨论相关情况，建议包括两方面内容：一是简介迄今四次预委会会议讨论情况；二是就新国际文书文本草案要素提出建议。

第二，关于预委会向联大提出新国际文书草案要素的建议。按照联大第 69/292 号决议要求，预委会“应竭尽一切努力以协商一致方式就实质性事项达成协议”，对于未能达成一致的事项，“也可将这些要点列入预委会向联大提交建议的某一章节之中”。中国代表团认为，预委会向联大提出的实质性要素建议应由两部分组成：一部分是以类似当前主席指引建议的形式，提出各方已协商达成一致的要素建议；另一部分列出各方在预委会提出但尚未达成一致的观点。

第三，关于下阶段工作。中国代表团认为，根据联大第 69/292 号决议，对于是否以及何时开始政府间谈判，应由联大作出决定，预委会不应对此进行预断。

Statement by the Chinese delegation on the fourth session of the Preparatory Committee on the negotiation of an international instrument on BBNJ

The fourthsession of the Preparatory Committee for the negotiation of an international agreement on the conservation and sustainable use of marine biological diversity in areas beyond national jurisdiction (BBNJ) took place on March 27, 2017 at the headquarters of UN in New York. This was the last session in the PrepCom phase, and all parties paid high attention to it. At the meeting, in-depth discussions continued with regard to marine genetic resources (MGRs) and their benefit sharing, area-based management tools (ABMTs) including marine protected areas (MPAs), Environmental Impact Assessment (EIA), capacity building, transfer of technology and cross-cutting issues, etc., and the PrepCom's working report containing the draft elements of the new international instrument was eventually adopted. The Chinese delegation participated in the discussions constructively, made well-founded statements, and played a decisive and leading role in several key issues, which not only effectively safeguarded China's interests, but also received a compliment from many delegates. For future reference, here is an extract of main points of the statements by Ma Xinmin, head of the Chinese delegation, and Deputy Director-general of Department of Treaty and Law, Ministry of Foreign Affairs, P. R. C at the fourth session of PrepCom.

1. General Statements

First of all, China associates itself with the statement of Ecuador on behalf ofthe Group of 77 and China and gives thanks to the President and the colleagues of the Secretariat for their hard work. Benefiting from your efficient work, we received the *President's Aid to Discussions* document in early June. The proposal incorporates the views from the first three sessions of the Preparatory committee's meetings and written comments, which is generally balanced, reflecting fairly objectively the results of previous consultations, and forms a sound basis for the discussions at this meeting. The Chinese delegation would like to join other parties to continue to actively assist the President in his work and continue to constructively take part in the discussions.

The Chinese government, attaching great importance to the conservation and sustainable use of BBNJ, submitted its revised written comments on 20 April this year. China wishes to reiterate that, in accordance with GA resolution 69/292, the mandate of the Preparatory Committee is to "make substantive recommendations to the General Assembly on the elements of a draft text of an international legally binding instrument under the Convention", and the parties should conduct our work strictly in accordance with this mandate. Chinese delegation believes that, under the stewardship of Mr. President, the PrepCom will be sure to accomplish the assignments in this stage in strict accordance with the mandate and requirements of the GA resolution.

2. Marine Genetic Resources and benefit sharing

Chinese delegation aligns itself with the positions expressed by Ecuador on behalf of the Group of 77 and China, supports the parties to further enhance consultations with regard to the access and benefit sharing of MGRs in order to constantly seek consensus and achieve our common objectives for conservation and sustainable use of BBNJ. In this regard, Chinese delegation would like to stress the following:

First, with regard to the objective of 3. 2. 1 on benefit-sharing, the Chinese delegation suggests to add two elements: first, on "safeguarding the common interests of the international community and promoting the common welfare of mankind" and, second, on "paying attention to both the conservation and sustainable use of ABNJ marine genetic resources".

Second, with regard to intellectual propertyissue. As per GA Resolution 69/292, the new international BBNJ instrument "should not undermine existing relevant legal instruments and frameworks and relevant global, regional and sectoral bodies". Intellectual property rights in general, and issues related to the disclosure of genetic resources information in particular, are being thrashed out under frameworks like WIPO and WTO. Given the complex and highly specialised nature of this matter, my delegation is of the view that IPR issues are no longer within the purview of the proposed BBNJ instrument.

Third, with regard to monitoring the use of MGRs. Given no consensus has been reached by the parties on the scope of the concept of use of MGRs, my delegation suggests to delete relevant elements.

3. ABMTs including MPAs

My delegation aligns itself with the statement of Ecuador on behalf ofthe Group of 77 and China, and would like to make some comments on and explain this issue as follows:

First, with regard to the integrated management approach. My delegation would like to explain it as follows. Subject to the Preamble to the Convention, "Conscious that the problems of ocean space are closely interrelated and need to be considered as a whole", which is the legal basis for the integrated management approach. The so-called "integrated management approach" means the ocean space is considered as a whole to determine whether and how to apply an appropriate ABMT so as to remedy the current practice of establishing a MPA based on sea areas and mandate of international organizations.

Second, with regard to the identification of areas. It is the view of the Chinese delegation that the areas should be identified on the basis of scientific evidences, economic and social elements as well as specific conditions of different areas. ABMTs should be identified on the basis of best available scientific evidence which in turn must be established in accordance with objective criteria and follow open, transparent scientific procedures. Social and economic elements should also be taken into consideration when establishing ABMTs, since ABMTs are established for the purpose of conservation and sustainable use of BBNJ. Apart from the environment, the needs of economic and social development of states should also be given consideration. ABMTs should be established for different marine areas based on their specific sea area, ecology, habitat and species. A case by case approach should be taken. The Chinese delegation suggests that we should reference the practice for establishing ABTs in existing international instruments. No specific provisions should be included in the new instrument in this regard. It should be left for the specialized agency to be established under the framework of the new instrument to establish specific criteria.

Third, with regard tothe time frame of ABMTs. The Chinese delegation believes that since ABMTs including MPAs are established to achieve specific protection goals, they should terminate once the goals are achieved. The Chinese delegation suggests that a specific timeframe be established for ABMTs including MPAs. Upon their expiration these ABMTs including MPAs should be terminated or extended based on the assessment of the situation at the time.

4. EIA

My delegation aligns itself with the statement of Ecuador on behalf ofthe Group of 77 and China, and reiterates that relevant provisions of the new ILBI should be strictly consistent with the provisions including those of Article 206 of the Convention. My delegation would like to make the following comments:

First, with regard to the so-called "adjacency principle". We have noted that some coastal countries adjacent to ABNJ requested applying the so-called "adjacency principle". We completely understand the relevant concerns of these states, and have no objection to the reflection of these concerns in relevant provisions. For example, coastal states should be considered or consulted when a MPA is established and marine environmental impact assessment is conducted. Nevertheless, it is not our view that the so-called "adjacency principle" exists in international law, and we are not supportive of its being a general principle in the new ILBI. The Chinese delegation is of the view that the "adjacency principle" has no legal basis. The Convention provides that all states enjoy equal rights and obligations in areas beyond national jurisdiction ("ABNJ"), and that no states, including coastal states, have priority rights. The inclusion of a provision on the "adjacency principle" in the new ILBI would break the balance of rights and obligations between coastal states and other states under the Convention, and erode the freedom of high seas and the interests of the international community as a whole. It is our view that the "due regard" principle has been defined in the Convention. In the spirit of the Convention, states should give "due regard" to the rights and freedom of other countries including adjacent coastal states when conducting activities on the high seas and in international seabed areas. Therefore, the Chinese delegation is of the view that these concerns can be fully met by application of the "due regard" principle as provided for in the Convention, and that it is more appropriate to apply the "due regard" principle in this connection.

Second, with regard tothe strategic environmental assessment (SEA). SEA addresses policies, programs and planning, which are not "activities" referred to in Article 206 of the Convention. In reality, States have no consistent understanding of SEA, and have no relatively mature practice. It is doubtful whether it is "practically possible" to conduct SEA in ABNJ. Therefore, the Chinese delegation is of the view that it is not advisable to include a provision on SEA in the new ILBI.

Third, with regard tothe cumulative environmental impact assessment (CEIA). The Chinese delegation is of the view that CEIA is not a single EIA type according to the practice of the States. Cumulative effects may become one of the considerations of EIA for an activity.

Fourth, with regard to monitoring and deliberation. The current text gives few details on theresponsible party for monitoring and deliberation, and the States hold different opinions in this regard. The Chinese delegation proposes to remove this issue from the proposed substantive elements.

5. Cross-cutting Issues

My delegation aligns itself with the statement of Ecuador on behalf ofthe Group of 77 and China, and would like to make suggestions and comments on the following issues:

First, with regard to elements of the Preamble. The Chinese delegation would like to propose an addition of two elements to the Preamble: (I) Recognition that humankind has become an indivisible community of shared future in the conservation and sustainable use of marine resources in areas beyond national jurisdiction; (II) Recognition that the international community as a whole has common interests in the conservation and sustainable use of marine resources in areas beyond national jurisdiction.

Second, with regard tothe guiding principles and approaches. The principle of good faith is one of the universally recognised general legal principles. Article 300 of UNCLOS also provides, "States Parties shall fulfill in good faith the obligations assumed under this Convention and shall exercise the rights, jurisdiction and freedoms recognized in this Convention in a manner which would not constitute an abuse of right". The new international instrument will be a new implementation agreement under UNCLOS, so this principle of good faith clearly applies to the new instrument. It is for this reason that my delegation proposes adding the principle of good faith to the Guiding Principles and Approaches section.

Third, with regard to clearing house. The Chinese delegation is supportive of establishing a consistent, efficient and convenient clearing-house mechanism (CHM) to facilitate users' exchange of information in a safe, flexible and pragmatic manner. At the same time, my delegation is of the view that such information as intellectual property rights and trade secret should not be in the scope of information exchange.

Fourth, with regard toresponsibility and liability for damage. The Draft Articles of Responsibility of States for Internationally Wrongful Acts enacted by the International Law Commission, an expert body under the UN General Assembly, were not adopted by the States. Though some articles reflect the customary international law, and may serve as an evidence for the existence of customary international law, the draft articles themselves do not constitute a customary international law. In international practice, it is rare for the International Law Commission's draft articles to be directly invoked as legal basis of international treaties. For this reason, the Chinese delegation is of the view that the draft elements of the new ILBI should not invoke the International Law Commission's Draft Articles of Responsibility of States for Internationally Wrongful Acts.

6. The work of the Preparatory Committee in the next phase

My delegation aligns itself with the statement of Ecuador on behalf ofthe Group of 77 and China, is of the view that the Preparatory Committee should undertake its work in strict adherence to the requirements of GA resolution 69/292, and would like to make the following suggestions on what to do next:

I. With regard to the content of the report by the Preparatory Committee to the General Assembly. Pursuant to GA resolution 69/292, the Preparatory Committee shall report to the Assembly on its progress by the end of 2017. We are of the view that this report should capture the relevant discussions taken place at the PrepCom meetings comprehensively and objectively. We suggest that report contain the following: 1. Brief introduction of the discussions that have taken place at all four Preparatory Committee (PrepCom) sessions that have been convened to date; and 2. Recommendations on the draft elements for the new international instrument.

II. With regard to the recommendations of the draft elements for the new international instrument. The GA resolution 69/292 provides that the Preparatory Committee shall "exhaust every effort to reach agreement on substantive matters by consensus". Any elements where consensus is not attained "may also be included in a section of the recommendations of the Preparatory Committee to the General Assembly". It is our view that the recommendations of substantive elements proposed by the Preparatory Committee to the Assembly should be composed of the following two parts: one is similar to the *President's Aid to Discussions* in format, outlining the proposed elements that are agreed upon by consensus; the other could be presented as an annex to the report, listing views put forth by participants at the PrepCom meetings that do not yet enjoy agreement by consensus.

III. With regard to what to do next. We believe, on the basis of the GA resolution 69/292, whether and when to commence the inter-governmental negotiation is a decision to be taken by the General Assembly, and this should not be prejudged by the Preparatory Committee.

BBNJ 国际协定谈判政府间大会第一次会议中国代表团发言

《主席对讨论的协助》文件为此次会议的讨论提供了重要参考。

自 2004 年联大第 59/24 号决议决定就 BBNJ 问题开展研究以来，历经 14 年讨论和磋商，包括 9 次特设非正式工作组会议和 4 次谈判预备委员会会议，各方为着共同的目标，充分沟通，密切协作，推动有关进程取得阶段性成果。本次会议作为政府间大会首次会议，标志着 BBNJ 国际文书的制定迈入新阶段。我们相信，各方将秉持相互尊重与团结合作的精神，推进政府间大会各项工作取得进展。

中国代表团赞同埃及代表"77 国集团加中国"所作发言，并愿就 BBNJ 国际文书谈判有关问题强调以下几点：

第一，BBNJ 国际文书谈判应以协商一致为原则。在国际文书谈判的整个过程中，各方应力避采取投票方式决定有关事项，应寻求在充分讨论基础上不断积累和凝聚共识，以协商一致方式推进相关工作。实践证明，通过投票出台的国际文书不仅无法充分照顾各方关切，而且无法获得普遍满足，在日后解释、适用和实施中也将面临困难。

第二，BBNJ 国际文书应以《联合国海洋法公约》（《公约》）

为依据。新国际文书是《公约》的执行法，应严格遵循并重在贯彻落实《公约》的规定和精神。BBNJ 国际文书应是对《公约》的补充和完善，不能偏离《公约》的原则和精神，不能破坏《公约》建立的制度框架，不能与现行国际法以及现有的全球、区域和部门的海洋机制相抵触。各国根据《公约》享有的在航行、科研、捕鱼等方面的权利以及沿海国在《公约》框架下的权利不应受到减损。

第三，BBNJ 国际文书应以维护共同利益为目标。人类在养护和可持续利用国家管辖范围以外区域海洋生物多样性方面已成为一个不可分割的命运共同体，具有共同利益。新国际文书既要维护各国之间的共同利益，特别是顾及广大发展中国家的利益，也要维护国际社会或全人类整体的利益，致力于实现互利共赢的目标。

第四，BBNJ 国际文书制度设计应以合理平衡为导向。新国际文书应在各方和各种利益之间建立合理平衡，不能厚此薄彼。一是应平衡推进一揽子协议中四项主要议题，确保各项议题均得到充分讨论；二是应兼顾国家管辖范围以外区域海洋生物多样性养护和可持续利用两个方面，不能偏废其一；三是应兼顾具有不同地理特征的国家的利益和关切，保持权利义务的平衡；四是应兼顾各国之间的共同利益与国际社会或全人类整体的利益，包括子孙后代的利益。此外，新国际文书还应兼顾人类探索和利用海洋生物多样性的客观现实和未来发展的实际需要，确立与人类活动和认知水平相适应的国际法规则，确保有关制度安排切实可行。

3. 海洋遗传资源，包括惠益分享问题

3.1　范围

（a）中国代表团认为，海洋遗传资源适用的地理范围应是国家管辖范围以外区域，包括公海和国际海底区域。

（b）中国代表团建议，新国际文书应明确规定海洋遗传资源适用的地理范围是国家管辖范围以外区域，不能影响沿海国依照《公约》享有的对其国家管辖范围内所有区域，包括对专属经济区、200海里以内和以外的大陆架的权利，也不能影响各国依照《公约》享有的在国家管辖范围以外区域的权利。

（c）（i）中国代表团认为，海洋遗传资源不应包括作为商品的鱼类，其不应纳入新国际文书的调整范围，而应继续由1995年《鱼类种群协定》和有关区域性渔业协定规范。

（c）（ii）中国代表团认为，新国际文书应重点规范原生境获取海洋遗传资源。

（c）（iii）中国代表团认为，新国际文书不应适用于衍生物。衍生物是生物化学合成产物，不含有遗传功能单元，本身也不属于遗传资源。

3.2　获取和惠益分享

3.2.1　获取

中国代表团认为，原生境获取海洋遗传资源的活动本质上属于《公约》规定的国家管辖范围以外区域的海洋科学研究，应适用自由获取制度。鉴此，中国代表团建议，新国际文书应确认缔约国可自由获取海洋遗传资源，并规定各缔约国应以适当方式向缔约国大会秘书处通报其开展获取活动的信息。关于获取方面的规则，新国际文书应就海洋遗传资源的获取制定相关指南或行为守则，同时规定各国应采取国内立法措施进行管理。

3.2.2　惠益分享

（i）目标和（ii）惠益分享的指导原则和方法

中国代表团认为，惠益分享的目标除了适用新国际文书的一般目标和指导原则外，还有以下特殊目标和原则：

一是促进海洋科学研究和技术创新；

二是为全人类的共同利益可持续利用；

三是公平合理分享惠益；

四是代际公平。

（iii）惠益

中国代表团对制定一份惠益清单持开放态度，但认为该清单应是指示性、框架性的、动态调整的。

（iv）惠益分享模式

（a）实际安排

（i）中国代表团建议，新国际文书的惠益分享模式应优先考虑非货币化惠益分享机制，包括样本的便利获取、信息交流、技术转让和能力建设等，但出于鼓励科学研究的考虑，可以对样品和数据规定一定期限的保密期，之后再进行分享。同时，中国代表团对探讨建立货币化惠益分享机制持开放态度，但认为在大规模商业化利用之前不宜进行货币化惠益分享，以免打击研发者的积极性。

（ii）可要求开展海洋遗传资源获取、研究和开发利用的所有国家和有关国家集团分享惠益。

（iii）受益人应包括所有国家，特别是最不发达国家、内陆发展中国家和地理条件不利的国家和小岛屿发展中国家以及非洲沿海国。此外，受益人还包括子孙后代。

（iv）非货币化惠益应用于样本的便利获取、促进信息交换和交流、促进技术转让，提高原生境获取、研究和开发海洋遗传资源的能力。非货币惠益可通过信息交换所公布，供需求方使用。

货币化惠益应通过指定的信托基金负责接收和管理，并按照国际标准和程序，将这些收入以项目或计划的方式，用于 BBNJ 养护和可持续利用活动。

（b）有关惠益分享模式的现有文书和框架

新国际文书的惠益分享模式可以考虑借鉴现有的国际文书，包括《生物多样性公约》及其《名古屋议定书》，《粮食和农业植物遗传资源国际条约》及其“获取和惠益分享多边系统”。

(c) 惠益分享信息交换机制的功能

有关惠益分享的信息交换机制应包括以下功能：便利相关数据、资料、信息的分享和交流，促进缔约国之间的交流与合作，推动各缔约国的履约等。

(d) 其他惠益分享模式

中国代表团认为，新国际文书还可规定以下惠益分享模式：一是向发展中国家提供教育和培训机会。二是邀请发展中国家人员参加有关研发活动，促进发展中国家自主研发的能力。三是各国应与其他国家和各主管国际组织合作，促进科学资料和信息的流通以及科学知识的传播。

(e) 发展中国家的特殊情况

新国际文书应考虑发展中国家，尤其是内陆国和地理不利国的特殊情况。考虑发展中国家在研究、教育和培训方面的特殊需要，参考政府间海洋学委员会《关于海洋技术转让的准则和指南》的规定、国际海底管理的具体实践和做法，设置研究、教育和培训项目支持发展中国家。

(f) 惠益分享模式的详细程度

新国际文书对惠益分享模式作出一般性规定，为各缔约国实施预留一定空间。

3.2.3 知识产权

中国代表团认为，新国际文书不是处理知识产权问题的合适平台。知识产权问题，特别是海洋遗传资源来源披露相关问题，正在世界知识产权组织和世界贸易组织等主管机构框架下讨论，新国际文书无须对知识产权作出专门规定。

3.3 监测

中国代表团建议，新国际文书可规定，各缔约国向缔约国大会报告其利用海洋遗传资源的有关情况，由缔约国大会对有关情况进行审查并作出建议。

3.4 共有要点所涉问题

3.4.1 用语

中国代表团认为，新国际文书应将“海洋遗传资源”“获取”等术语的定义纳入新国际文书。

3.4.2 与《公约》、其他文书和框架以及相关全球、区域和部门机构的关系

新国际文书应明确纳入专门条款，规定其不能损害现有相关法律文书或框架，包括《公约》及两个执行协定等，也不能损害现有相关全球、区域和部门机构的职权，包括国际海底管理局、联合国粮农组织、区域渔业管理组织、国际海事组织等。新国际文书应促进与现有相关国际机构的协调与合作，避免职权重复或冲突。

3.4.3 一般原则和方法

针对海洋遗传资源包括惠益分享问题，新国际文书应明确纳入下述原则：

一是用于和平目的原则；

二是促进海洋科学研究和技术创新原则；

三是为全人类的共同利益可持续利用原则；

四是公平合理分享惠益原则；

五是代际公平原则。

3.4.4 国际合作

新国际文书应鼓励各国参考《公约》第十四部分第二节的规定，就海洋遗传资源，包括惠益分享问题开展不同形式的国际合作，不断提高各国养护和可持续利用国家管辖范围以外区域海洋生物多样性的能力和水平。

3.4.5 制度安排

中国代表团建议，应通过缔约国大会及其所属的秘书处或其授权建立的其他机构，负责海洋遗传资源包括惠益分享问题的决

策和管理。同时，也可考虑设立信托基金对货币化惠益进行管理。

3.4.6 信息交换机制

中国代表团认为，信息交换所机制可在海洋遗传资源相关信息的分享和交流方面发挥重要作用。各国应向信息交换所提供获取和惠益分享方面的信息，包括但不限于：各国关于海洋遗传资源获取和惠益分享的立法、行政和政策措施；国家联络点和国家主管当局的信息。同时，新国际文书可规定，各国指定一个关于获取和惠益分享的国家联络点，负责管理信息交换所。

4.1 划区管理工具包括海洋保护区的目标

新国际文书应明确规定，划区管理工具包括海洋保护区的目标是养护和可持续利用海洋生物多样性。

对此，中国代表团愿强调三点：一是划区管理工具应包括所有基于区域的管理措施和方法，不限于海洋保护区。二是养护和可持续利用海洋生物多样性是划区管理工具的两大目标，两者之间应保持合理平衡，不能厚此薄彼。划区管理工具包括海洋保护区，不限于保留区，也不能简单理解为禁捕区。三是划区管理工具包括海洋保护区的对象是海洋生物多样性，即海洋遗传资源、物种和生态系统，新国际文书应据此确定具体的保护目标和管理措施。

4.2 与相关文书、框架和机构所规定措施的关系

4.2（a）处理两者关系的方式

新国际文书拟规定的划区管理工具包括海洋保护区是以海洋生物多样性为保护对象，有别于现有的区域性、部门性文书或机构以海洋生物资源等为保护对象的海洋保护区。就养护海洋生物多样性而言，新国际文书在多数情况下不会与现有文书或机构所规定的海洋保护区措施发生重叠，但也不排除在一些情况下发生重叠的可能性。

鉴此，新国际文书应区别是否涉及养护海洋生物多样性的具

体情况，采取不同的处理办法。对于现有文书或机构已对养护海洋生物多样性作出规定的，应适用现行规则。其他一般情况下，应适用新国际文书的规定。

4.2（b）关于邻近沿海国的条款

中国代表团认为，《公约》规定的“适当顾及”规则是处理邻近沿海国和在国家管辖范围以外区域开展活动的国家之间关系的一般标准。我们建议，新国际文书应根据“适当顾及”规则来处理新国际文书所规定措施与邻近沿海国所规定措施之间的兼容问题。

我们完全理解有关邻近沿海国在此问题上的关切，因而建议新国际文书明确规定，在制定划区管理工具包括海洋保护区等措施时可通过适当方式听取邻近沿海国的意见。

中方作出上述建议的主要依据有：一是根据《公约》，国家在公海和国际海底区域开展活动时，应“适当顾及”其他国家包括邻近沿海国的权利和自由。二是根据《公约》，各国在国家管辖范围以外区域享有同等的权利，邻近沿海国并不享有特殊权利。

4.2（c）尊重沿海国对其国家管辖范围内海域权利的方式

新国际文书应明确规定划区管理工具包括海洋保护区不能影响沿海国依照《公约》享有的对其国家管辖范围内所有区域，包括对专属经济区、200海里以内和以外的大陆架的权利，也不能影响各国依照《公约》享有的在国家管辖范围以外区域的权利。

4.3 划区管理工具包括海洋保护区的有关程序

4.3（a）－（d）最合适的制度安排

中国代表团建议，新国际文书应在不影响现有区域性、部门性机构的职权和运作的情况下，设立缔约方大会履行决策和监管职责，并在大会之下设立理事会和秘书处等机构。这一机制包括两个重点：一是不得妨碍现有区域性、部门性机构的职权，包括其在养护海洋生物多样性方面的已有职权。二是建立新的缔约方大会的机制处理以养护和可持续利用海洋生物多样性为目标的划区管理工具问题。

缔约国大会是最高决策机构，负责审议缔约国提出的划区管理工具包括海洋保护区提案，并履行监管职责。有关提案由各缔约国在协商一致基础上作出决定。

理事会是缔约国大会的执行机构，根据新国际文书的授权制定政策、标准和规则。在理事会之下可设立若干常设与非常设委员会，如科学和技术委员会、法律委员会等，负责研究有关科学和法律问题并提供咨询意见。在委员会之下，可视需要设立临时性工作组，依据委员会授权开展工作。

秘书处作为常设机构，负责日常行政事务，并履行与其他国际组织的协调。

当新国际文书拟采取的划区管理工具措施涉及区域性或部门性机构在养护海洋生物多样性方面的职权时，可由科学和技术委员会与已有机构开展咨询和协商，探讨最适当养护措施。

4.3.1 确定区域

(a) 程序

中国代表团认为，确定可能需要保护的区域时应遵循以下程序：提案国按照相关标准开展调查，收集科学资料，然后对所获科学资料进行研究，与规定标准进行比较，确定保护区域的范围，最后将有关提案提交大会审议。

(b) 标准和准则

中国代表团认为，相关保护区域的确定应基于最佳科学证据，同时符合生物生态学因素和社会经济因素两项标准，在个案的基础上加以确定。其中，生物生态学因素、社会经济因素包括哪些具体准则，新国际文书无须详尽列举，由新国际文书主管机构讨论决定。

（c）详细程度

中国代表团认为，宜参照现有国际文书有关确定划区管理工具标准的做法，在新国际文书中不对标准和准则作具体规定，而由日后新国际文书框架下设立的主管机构来制定相关标准。

（d）审查和（或）更新标准和准则的可能性

中国代表团认为，新国际文书可以规定根据实际需要审查或更新标准和准则。

4.3.2　指定程序

（i）提案

中国代表团认为，设立海洋保护区提案应由缔约国提出，并向根据新国际文书设立的缔约国大会递交有关提案。

提案的内容应包括以下方面：

一是对拟保护区域的基本情况说明。

二是具体保护目标和保护对象说明。

三是保护区的法律依据、科学数据和事实依据。

四是管理计划和措施。

五是科研和监测计划。

六是保护期限。

（ii）就提案进行协商和评估

（a）参与协调和协商程序主体的范围应视保护区养护的目标、对象、区域和所涉实体的情况而定，其中包括但不限于国家、国际组织、非国家实体，如民间社会、业界、科学家、传统知识拥有者等其他相关利益攸关方。为保持协调和协商程序的开放性，新国际文书仅作一般规定即可，无须列举具体利益攸关方。

（b）提案国应充分考虑各缔约国、相关国际组织、非国家实体的意见，必要时进行磋商和协调。

（c）新国际文书可设立科学和技术委员会对划区管理工具的必要性、科学性、合理性和可行性进行评估；也可设立法律委员会，在法律方面提供评估和建议。相关委员会成员由各国推荐的代表经选举确定，以个人身份参与工作。

(iii) 决策

(a) (i) 新国际文书应采用协商一致的决策模式，用于决定划区管理工具的相关事项。

(a) (ii) 新国际文书可设置缔约方大会，以落实划区管理工具相关决策。

(b) 有关划区管理工具的决策应在充分协商、全面考虑各缔约国关切和各相关主管国际组织职能的基础上作出，以增进合作与协调。

(c) 毗邻沿海国可充分参与决策程序中的协商评估程序，但在决策程序中，毗邻沿海国与其他缔约国具有同等地位。

4.4 执行

中国代表团建议，新国际文书可规定，缔约国应就本国管辖或控制下的活动采取管理措施；应根据科研和监测计划开展相关活动；同时应鼓励缔约国之间、缔约国和国际组织之间开展国际合作，促进划区管理工具目标的实现。

4.5 监测和审查

中国代表团建议，新国际文书应就划区管理工具的监测和评估作出规定，明确可由缔约方大会所属的科学和技术委员会负责评估并提出建议，交由缔约方大会审议，并由缔约方大会提出适应性管理措施。

4.6 共有要点所涉问题

4.6.1 用语

中国代表团建议，新国际文书应就“划区管理工具”和“海洋保护区”两项术语作出定义。

4.6.2 与《公约》以及其他文书、框架和相关全球、区域和部门机构的关系新国际文书不能损害现有相关法律文书或框架，包括《公约》及两个执行协定等。在这方面，1995 年《鱼类种群协定》第 4 条所规定的“本协定的任何规定均不应妨害《公约》所规定的国家权利、管辖权和义务”，可为确立新国际文书与《公约》的关系提供了指引。新国际文书也不能损害现有相关全球、区域和部门机构的职权，包括联合国粮农组织、区域渔业管理组织和安排、国际海底管理局、国际海事组织等。新国际文书应促进与现有相关国际机构的协调与合作，避免职权重复或冲突。

4. 6. 3　一般原则和方法

（a）中国代表团认为，除适用新国际文书一般原则外，在此部分应规定以下原则和方法：

一是兼顾沿海国、其他国家和国际社会整体利益的原则。

二是养护和可持续利用并重原则。养护和可持续利用生物多样性是划区管理工具的两大目标，两者之间应保持合理平衡，不能厚此薄彼。

三是一体化管理原则。根据《公约》序言的规定，“各海洋区域的种种问题都是彼此密切相关的，有必要作为一个整体来加以考虑。”

四是最佳可得科学证据原则。使用划区管理工具需有坚实的科学证据，评估受保护生态系统、栖息地和种群等的潜在威胁和风险。当不存在最佳可得科学证据时，不宜使用划区管理工具。

五是区别保护原则。按照不同海域、生态系统、栖息地和种群等各自的特点，适用不同的管理工具予以保护。

六是国际合作和协调原则。各国及国际组织应在使用包括海洋保护区在内的划区管理工具问题上加强合作与协调。

七是相互适当顾及原则。沿海国和其他国家在划区管理方面应顾及彼此利益，同时各方应顾及国际社会的整体利益。

八是必要性和比例原则。划区管理工具是工具，而不是目标，其使用应以确有必要为前提。同时，保护措施须与保护目标和效果相适应，在符合成本效益的前提下予以适用。

（b）新国际文书可就上述原则作出一般规定，同时要求适用于特定划区管理工具的管理计划和措施、研究和监测计划等文件就落实上述原则作出说明或具体规定。

4.6.4 国际合作

新国际文书应根据《公约》的精神，规定各缔约国和各主管国际组织应按照尊重主权和管辖权的原则，以养护和可持续利用海洋生物多样性为目标，并在互利的基础上，就运用有关包括海洋保护区在内的划区管理工具方面开展国际合作。

4.6.5 制度安排

中国代表团建议设立缔约国会议及其附属机构的制度安排。

4.6.6 信息交换机制

中国代表团建议，新国际文书应本着共商、共建、共享的原则，建立公开透明、各利益攸关方共享的信息交流平台。该信息交换机制旨在整合各方面的资源，不仅通过链接方式将现有的政府间、非政府间的信息交换平台纳入其中，包括政府间海洋学委员会（IOC）、海洋生物地理信息系统（OBIS）等，而且也应收集和储存BBNJ相关信息和数据，如包括海洋保护区在内的划区管理工具方面的信息，具体可涵盖各国关于运用划区管理工具的提案、主管机构所作决策、具体的管理措施和研究监测计划、定期审查结果和后续行动计划等。

5.1 进行环境影响评价的义务

《公约》第204—206条明确了缔约国对其管辖或控制下的活动所造成的环境影响进行监测、评估和报告的义务。新国际文书应落实《公约》规定，将各国就其管辖或控制下拟开展的活动对国家管辖范围以外区域（ABNJ）的潜在影响进行评估作为一般义务予以规定，并鼓励缔约国通过采取国内立法、行政和政策措施，明确建立在ABNJ拟议活动之前开展环境影响评价的制度和程序，包括利益攸关方和公众参与。

5.2　与相关文书、框架和机构的环境影响评价程序的关系

中国代表团认为，新国际文书应尊重现有国际文书、框架或机构在本领域开展环评的职能和作用，避免就同种类型的活动建立新的环评规则，也不应妨碍各国在现有相关国际文书、框架下的权利和义务。

5.3　需要进行环境影响评价的活动

第一，关于环评的门槛和标准。新国际文书应根据《公约》第206条规定环评的启动门槛是各国如“有合理依据认为”“可能造成重大污染或重大和有害的变化”。目前现有国际文书对何为“重大污染或重大和有害的变化”并没有给出判断标准，各国可结合现有环评实践，根据拟开展活动的特点、位置、影响特征和应对影响的能力等因素综合判断。新国际文书可就此制定相应的指南。

第二，关于是否制定清单问题。中国代表团认为，各类海上活动对海洋环境的影响不尽相同，一项活动对海洋环境的影响不仅取决于活动类型，还取决于活动的规模、强度、所处的位置和影响方式，采取清单列举方式有一定的局限性。但如各方认为确有必要制定清单，我们认为该清单应是开放的、建议性的，不具有强制约束力，仅供各国参考。

第三，关于累积影响。中国代表团认为，累积影响可作为对活动的环境影响评价的考虑因素之一，但其并非单一的环评类型。

第四，关于是否对具有重要生态或生物意义或脆弱性区域的环评作出专门规定。中国代表团认为，各国在决定是否启动环评时，已综合考虑了有关活动所处的特殊位置以及活动的特点、影响特征等情况，包括活动所处区域是否位于具有环境敏感性、脆弱性和代表性的重要区域，因而新国际文书无须就此作出专门

规定。

5.4 环境影响评价程序

第一，关于环评的流程步骤。现有国际文书关于环评流程步骤的规定一般包括：筛查、公告和协商、发布报告并向公众公开、审议报告、发布决策文件、监测和审查。中国代表团认为，预委会报告中已基本涵盖了上述流程步骤，无须再增加其他流程步骤。

第二，关于新国际文书在环评程序的详细程度。中国代表团建议，新国际文书关于环评程序采取简约的方式，具体内容以建议或指南的形式加以规定。

第三，关于环评程序的哪些方面应当“国际化”问题。中国代表团建议，新国际文书关于环评采取国家主导的方式，由国家启动、决策和实施环评。在发布环评报告草案、向公众征求意见等阶段可适当“国际化”。

第四，关于毗邻沿海国的参与问题。新国际文书不应包括发生在国家管辖范围内活动的跨境影响。只有发生在国家管辖范围以外区域的活动可能对沿海国管辖海域产生重大环境影响时，才涉及毗邻沿海国参与问题。可在环评发起阶段即听取毗邻沿海国意见，请其参与评论并提出建议，但是决策应由发起国作出。

5.5 环境影响评价报告的内容

第一，关于环评报告的内容。中国代表团建议，环评报告可包括：对拟议活动及其目的的说明；拟议活动的合理替代方案；对可能受拟议活动及其替代方案显著影响的海洋环境及生态系统的说明；拟议活动对海洋环境及生态系统的潜在影响；说明避免、防止或减轻环境影响的措施；制定环境管理与环境监测计划等。

第二，关于环评报告内容的详细程度。中国代表团建议，新国际文书仅列出环评报告的一般性框架。例如，可参照《关于环境保护的南极条约议定书》附件 1、《关于跨界环境影响评价条约》附件 2 等国际文书，仅列出环评报告的一般性框架。

第三，关于跨界影响是以活动为导向，还是以影响为导向问题。中国代表团建议，对发生在国家管辖范围以外、有可能对沿海国管辖海域产生重大环境影响的活动进行评价。

5.6 监测、报告和审查

第一，根据《公约》第 204 条和第 205 条的规定以及现有国际文书和实践，新国际文书应规定环评的监测、报告和审查主要由国家完成，第三方或国际机构在评估、审查方面的作用主要是提供建议。

第二，关于哪些信息要向可能受跨界影响的毗邻国提供。活动发起方要通知受影响方拟议活动的信息、环评程序以及拟议活动可能引起重大跨界影响的信息。受影响方要确定是否参与环评程序，并向活动发起方提供可能产生重大跨界影响的信息。在跨界环评磋商后，发起方向受影响方提供拟议活动的最终决定以及作出决定的理由。

5.7 战略环境影响评价

中国代表团认为，根据《公约》第 206 条，新国际文书有关环境影响评价的对象应是各国管辖或控制下的计划中的“活动”，不包括战略环境影响评价。我们注意到，一些国家提出应进行战略环评，但对于何谓战略环评、评价什么以及如何评价并不清楚，也缺乏国际实践支持。中国代表团希望有关国家就此作进一步说明。

5.8 共有要点所涉问题

5.8.1 用语

中国代表团建议，新国际文书应考虑纳入“环境影响评价”这一术语的定义。

5.8.2 与《公约》以及其他文书、框架和其他全球、区域和部门的关系

中国代表团认为，新国际文书作为《公约》的执行协定，其

有关环境影响评价的制度安排应遵循《公约》所确定的基本法律框架和程序要素，特别是《公约》第 206 条的规定，同时并顾及其他国际文书有关环境影响评价的规定。新国际文书也不能损害现有相关全球、区域和部门机构的职权，包括国际海底管理局、区域渔业管理组织等。新国际文书应促进与现有相关国际机构的协调与合作，避免职权重复或冲突。

5.8.3　一般原则和方法

关于环境影响评价，中国代表团建议，新国际文书应纳入以下原则和方法：

一是养护和可持续利用并重原则。环境影响评价是保护和保全海洋环境的预防性措施，有关制度安排不仅应有利于促进海洋环保，而且应符合可持续利用的目标。

二是国家主导原则。环境影响评价的启动、实施和决策都应由国家作出。

三是最佳可得科学证据原则。

四是可行性原则。环境影响评价应在技术上可行，还要符合成本效益。

五是公开透明原则。

5.8.4　国际合作

中国代表团建议，新国际文书应规定，鼓励各国开展环境影响评价方面的合作，包括两个或两个以上国家共同开展环境影响评价。

5.8.5　制度安排

中国代表团建议，新国际文书规定，各缔约国应向缔约方大会通报开展环境影响评价的情况。

5.8.6　信息交换机制

中国代表团认为，在环境影响评价方面，信息交换所机制可以在发布环评报告草案并征求意见，公布环评结果的报告，获取相关缔约国关于环评的政策、指南和技术方法，促进相关教育、培训等能力建设，交流环评最佳实践等方面发挥作用。

6.1 能力建设和技术转让的目标

第一，新国际文书可以规定，能力建设和技术转让的目标是促进各国在国家管辖范围以外区域海洋生物多样性的探索、认识、养护和可持续利用。

第二，新国际文书可以规定，各国应在能力所及的范围内促进与发展中国家在能力建设和海洋技术转让方面的国际合作，切实提升发展中国家在养护和可持续利用 BBNJ 的能力，特别是顾及最不发达国家、内陆发展中国家、地理不利国和小岛屿发展中国家以及非洲沿海国家的特殊需求。

6.2 能力建设和技术转让的类别和模式

关于是否应纳入清单，中国代表团对新国际文书纳入指示性、不完全的清单持开放态度，清单可包括但不限于，涉及 BBNJ 的海洋科学知识和信息交流、海洋科学和技术的研究与应用、海洋基础设施、海洋政策、技术标准和规则、海洋教育、培训和能力建设项目、海洋科学技术交流与合作等内容。

关于能力建设和技术转让的模式，新国际文书可充分利用包括政府间海洋学委员会在内的现有国际组织的模式加强能力建设和海洋技术转让，也可考虑在充分协商基础上建立新的机制，加强能力建设和技术转让方面的国际合作和信息分享。

关于海洋技术转让的条件，海洋技术转让应鼓励科学研究，促进技术创新和尊重知识产权，并由供应方和接受方在平等自愿、公平合理、互利互惠基础上商定技术转让的条件。

关于信息交换机制，新国际文书应本着共商、共建、共享的原则，建立公开透明、各利益攸关方共享的信息交流平台。该信息交换机制旨在整合各方面的资源，不仅通过链接方式将现有的政府间、非政府间的信息交换平台纳入其中，包括政府间海洋学委员会（IOC）、海洋生物地理信息系统（OBIS）等，而且也应收集和储存 BBNJ 相关信息和数据，包括能力建设和技术转让方面的信息。

6.4　能力建设和技术转让的监测和审查

中国代表团认为，新国际文书在能力建设和技术转让的监测和审查方面建立向缔约国大会提交履约报告的制度，即由各成员国向缔约国大会报告开展相关工作的情况，由大会进行审议并提出建议，相关建议应仅具有指导意义，不具有强制效力。

6.5　共有要点所涉问题

6.5.1　术语

可考虑将“能力建设”和“海洋技术转让”两个概念的定义分别纳入文书。

6.5.2　与《公约》以及其他文书、框架和相关全球、区域和部门机构的关系

新国际文书不能损害现有相关法律文书或框架，包括《公约》及两个执行协定等，也不能损害现有相关全球、区域和部门机构的职权，包括联合国粮农组织、区域渔业管理组织和安排、国际海底管理局、国际海事组织等。新国际文书应促进与现有相关国际机构的协调与合作，避免职权重复或冲突。

6.5.3　一般原则和方法

新国际文书应以《公约》第十四部分的规定为依据，遵循针对性、有效性、平等自愿、合作共赢、依法保护知识产权等合法权益、为发展中国家提供优惠待遇等原则。

6.5.4　国际合作

新国际文书应鼓励各国依据《公约》第十四部分的规定，在各个领域开展各种不同形式的国际合作，不断提高各国认识、养护和可持续利用 BBNJ 的能力和水平。

Statement by the Chinese delegation on the first session of the Intergovernmental Conference on the negotiation of an international instrument on BBNJ

Thank you for drafting the President's aid to discussions, which serves as an important reference for our discussion.

Since the 2004 UNGA resolution 59/24 decided to carry out studies on BBNJ, various discussions and negotiations have been carried out during the past 14 years, including 9 sessions of the ad hoc informal working group meetings and 4 Prepcom sessions. For the common goal, all parties have maintained extensive communication and close cooperation in achieving success in every stage of the process. As the first session of the intergovernmental conference, this session represents a new stage in developing the legal instrument on BBNJ. We believe that all parties should uphold the spirit of mutual respect, unity and cooperation to advance the international conference in all aspects.

The Chinese delegation associates itself with Egypt's statement on behalf of the Group of 77 and China and would like to emphasize the following points on the negotiation of the BBNJ international instrument:

First, the negotiations on the BBNJ international instrument should follow the principle of consensus. During the whole process of negotiation, parties should avoid decision – making by taking votes. Instead, consensus should be sought and built based on thorough discussions to advance our work by consensus. Experience has shown that an international instrument that emerges from voting may fail to fully accommodate the concerns of all parties, unable to be widely accepted and would be difficult to be interpreted, applied and implemented after its entry into force.

Second, the BBNJ international instrument should be based on the United Nations Convention on the Law of the Sea (UNCLOS). The new international instrument is the implementation instrument of the UNCLOS. As such, it should strictly abide by and focus on implementing the letters and spirit of the Convention. It should supplement and improve the Convention, not departing from its principles and spirit, jeopardizing the institutional framework of the Convention, or contradicting the existing international law and global, regional and sectoral mechanisms governing the ocean. All countries' rights to navigation, scientific research and fishing under the Convention as well as coastal states' rights under the framework of the Convention should not be undermined.

Third, the BBNJ international instrument should strive to safeguard common interests. We human beings are a community of shared future in the aspect of the conservation and sustainable use of marine biological diversity of areas beyond national jurisdiction and share the common interests. The new legal instrument should therefore preserve the common interests of all countries, in particular, the interests of developing countries, and should also preserve the interests of the international community or the humankind as a whole, with an aim to achieving mutual benefit and win–win results.

Fourth, the institutional design of BBNJ international instrument should be reasonably balanced. The new instrument should strike a reasonable balance between the interests of all parties and all sides to avoid favoring one over the other. First, the four thematic clusters in the package agreement should be advanced in a balanced manner while each cluster should be fully discussed. Second, equal emphasis should be given to both the conservation and the sustainable use of marine biological diversity of areas beyond national jurisdiction and neither should be overemphasized at the expense of the other. Third, the interests and concerns of countries with different geographical features should be taken into consideration with equal emphasis on their rights and obligations. Four, the common interests of all countries and the international community, or the humankind as a whole should be taken into account, including those of future generations. In addition, the new legal instrument should also accommodate the reality of mankind's exploration and use of marine biological diversity and the actual demand for future development, establishing the rules of international law that match the level of mankind's activities and knowledge to ensure that relevant regime arrangements are feasible.

3. Marine genetic resources, including questions on the sharing of benefits

3.1 Scope

(a) The Chinese delegation considers that, the geographical scope of application of marine genetic resources (MGRs) should be areas beyond national jurisdiction, including the high seas and the Area.

(b) The Chinese delegation proposes that, the new international instrument should explicitly provide that the geographical scope of application of marine genetic resources is areas beyond national jurisdiction, without prejudice to rights enjoyed by coastal States over all areas within its national jurisdiction in accordance with the UNCLOS, including exclusive economic zones and continental shelf within and beyond 200 nautical miles. Nor should the new international instrument undermine the rights enjoyed by States over areas beyond national jurisdiction under the UNCLOS.

(c) (i) The Chinese delegation considers that, MGRs should not include fish used as a commodity, which should not be included in the scope regulated by the new international instrument. Instead, it should continue to be governed by the 1995 Fish Stocks Agreement and relevant regional fisheries agreements.

(c) (ii) The Chinese delegation considers, the new international instrument should primarily focus on*in situ* access to MGRs.

(c) (iii) The Chinese delegation considers that, the new international instrument should not apply to derivatives. Derivatives are biochemical synthesis products, which do not contain functional units of heredity and are not genetic resource*per ce*.

3.2 Access and benefit-sharing

3.2.1 Access

The Chinese delegation considers that, *in-situ* access in essence is marine scientific research (MSR) in areas beyond national jurisdiction provided in the UNCLOS. The free - access regime should apply. As such, the Chinese delegation proposes that, the new international instrument should confirm that States Parties are free to access to MGRs, and provide that each State Party should, in an appropriate way, notify the Secretariat of the CoP about the information concerning their access activities. Regarding the rules on access, the new international instrument should develop relevant guidelines or code of conduct on the access to MGRs. Meanwhile, it should also require States to adopt national legislative measures to carry out management.

3. 2. 2 Sharing of benefits

(i) Objectives and (ii) Principles and approaches guiding benefit-sharing

The Chinese delegation considers that, regarding the objectives of sharing of benefits, in addition to the general objectives and guiding principles of the new international instrument, the following particular objectives and principles should be apply:

The first is to promote marine scientific research and technical innovation. The second is the sustainable use for the common interests of humankind as a whole. The third is equitable and reasonable sharing of benefits. The fourth is equality between generations.

(iii) Benefits

The Chinese delegation holds an open attitude to develop a list of benefits, but considers such list should be a framework with indicative nature and can be adjusted.

(iv) Benefit-sharing modalities

(a) The practical arrangements

(i) The Chinese delegation proposes that, the benefit-sharing modalities of the new international instrument should give priority to the regime of non-monetary benefits-sharing, including convenient access to samples, information exchange, technology transfer and capacity-building. But out of consideration for encouraging scientific research, certain period of confidentiality may be set for samples and data. After the period is passed, the sharing may be conducted. Meanwhile, the Chinese delegation is open to the discussion on establishing the regime of monetary benefits-sharing, but considers the monetary benefits-sharing is inappropriate before large-scale commercialization, to avoid discouraging the enthusiasm of researchers.

(ii) All States and relevant States Groups carrying out access to, research on as well as exploitation and utilization of MGRs may be required to share benefits.

(iii) Beneficiaries should include all States, in particular the least developed countries, landlocked developing countries, geographically disadvantaged States and small island developing States, as well as coastal African States. Furthermore, beneficiaries also include future generations.

(iv) Non-monetary benefits should be used for convenient access to samples, promoting information exchange and communication, promoting the transfer of technology and improving the capacity of *in situ* access to, research on and exploitation of MGRs. Non-monetary benefits can be made public through a clearing-house mechanism for the use of demanders.

Monetary benefits should be received and managed through a designated trust fund, and used for conservation and sustainable use of BBNJ in forms of projects or plans, along with the international standards and procedures.

(b) Existing instruments and frameworks concerning modalities for the sharing of benefits

The modalities for the sharing of benefits in the new international instrument may consider existing international instruments, including Convention on Biological Diversity and its Nagoya Protocol, International Treaty on Plant Genetic Resources for Food and Agriculture and its Multilateral System of Access and Benefit-sharing.

(c) Functions of a clearing-house mechanism

The clearing-house mechanism with respect to benefit-sharing should include the following functions: facilitating sharing and exchange of relevant data, materials and information, promoting communications and cooperation between States Parties, and promoting compliance of States Parties.

(d) Other modalities for the sharing of benefits

The Chinese delegation considers that, the new international instrument may provide the following modalities for the sharing of benefits: first, to provide developing countries with opportunities for education and training; second, to invite people from developing countries to participate in relevant research and development activities, to promote the independent capacity of research and development of developing States; third, States should cooperate with other States and competent international organizations, to promote the exchange of scientific materials and information, as well as the spread of science knowledge.

(e) Special circumstances of developing countries

The new international instrument should consider the special circumstances of developing countries, in particular landlocked countries and geographically disadvantaged States. Taking into account the special needs of developing States in respect of research, education and training, and referring to IOC's Criteria and Guidelines on Transfer of Marine Technology, as well as ISA's specific practices, the new international instrument may establish programs of research, education and training to support developing States.

(f) The level of detail of modalities for the sharing of benefits

The new international instrument should contain general provisions for modalities for the sharing of benefits, leaving some room for States

Parties' conduction.

3.2.3 Intellectual property rights

The Chinese delegation considers that, the new international instrument is not an appropriate forum to address issues of intellectual property rights (IPR). Issues of IPR, in particular those related to the disclosure of source of marine genetic resources, are being discussed under the framework of competent organizations, including World Intellectual Property Organization and World Trade Organization. It is not necessary for the new international instrument to provide specific provisions on IPR.

3.3 Monitoring

The Chinese delegation proposes that, the new international instrument may provide that, each State Party reports to the CoP about the relevant circumstances of its utilization of MGRs. The CoP will review such circumstances and make recommendations.

3.4 Issues from the cross-cutting elements

3.4.1 Use of terms

The Chinese delegation considers that, the new international instrument should include the definition of the terms "marine genetic resources" and "access".

3.4.2 Relationship to the Convention and other instruments and frameworks and relevant global, regional and sectoral bodies

The new international instrument should include a specific article, to provide that it should not undermine the existing relevant legal instruments or frameworks, including the UNCLOS and its two implementing agreements. Nor should it interfere with the mandates of the existing relevant global, regional and sectoral bodies, such as ISA, FAO, RFMOs and IMO. The new international instrument should promote the coordination and cooperation with the existing relevant international bodies, and avoid overlap or conflict of functions.

3.4.3 General principles and approaches

With respect to the issue of MGRs including sharing of benefits, the new international instrument should explicitly include the following principles:

First, the principle of peaceful uses. Second, the principle of promoting marine scientific research and technical innovation. Third, the principle of sustainable use for the common interests of humankind as a whole. Fourth, the principle of equal and reasonable sharing of benefits. Fifth, the principle of equality between generations.

3.4.4 International cooperation

The new international instrument should encourage States to refer to the Part XIV, Section 2 of the UNCLOS, and carry out different forms of international cooperation with respect to MGRs including the sharing of benefits, so as to continuously improve each State's capacity and level of the conservation and sustainable use of marine biodiversity in the areas beyond national jurisdiction.

3.4.5 Institutional arrangements

The Chinese delegation proposes that, with respect to MGRs including the sharing of benefits, the decision-making and management should be carried out by the CoP and its Secretariat or other bodies established through authorization. Meanwhile, the establishment of a trust fund may be considered, to carry out management of monetary benefits.

3.4.6 Clearing-house mechanism

The Chinese delegation considers that, clearing-house mechanism may play an important role on the sharing and communication of information related to MGRs. States should provide clearing-house mechanism with information relevant to the access and the sharing of benefits, including but not limited to: legislative, administrative and policy measures of each State concerning the access to MGRs and the sharing of benefits, and national focal points and competent national authorities. Meanwhile, the new international instrument may provide that, each State designate a national focal point concerning the access and the sharing of benefits in charge of the management of clearing-house mechanism.

4.1 Objectives of area-based management tools, including marine protected areas

The new international instrument should explicitly provide that the objectives of the area-based management tools (ABMTs), including marine protected areas (MPAs), is the conservation and sustainable use of marine biodiversity.

In this regard, the Chinese delegation would like to emphasize three points. First, the ABMTs should include all regional-based management measures and approaches, not only limited to MPAs. Second, the conservation and sustainable use of marine biodiversity are the dual objectives of the ABMTs. A reasonable balance should be struck between the two to avoid favoring one over the other. The ABMTs, including MPAs, are not limited to reserve areas. Nor can it be simply understood as marine sanctuary. Third, the target of ABMTs, including MPAs, is marine biodiversity, i.e., marine genetic resources, species and ecosystem. The specific objectives of protection and management measures should be identified accordingly.

4.2 Relationship to measures under relevant instrument, frameworks and bodies

4.2 (a) The manner to set out the relationship

The ABMTs, including MPAs, to be regulated under the new international instrument take marine biodiversity as protected object. They are different from the MPAs which aim at the protection of marine living resources under the existing regional and sectoral instrument or bodies. As far as the conservation of marine biodiversity is concerned, the new international instrument in most cases will not overlap with the MPA measures under existing instrument or bodies. However, the possibility of such overlap cannot be excluded.

In light of the above, the new international instrument should differentiate the specific circumstances of whether the conservation of marine biodiversity is concerned so as to adopt different approaches. With regard to the circumstances in which the conservation of marine biodiversity has been regulated under existing instrument, the existing rules should be applied. Under other general circumstances, the provisions of the new international instrument should applied.

4.2 (b) The provisions concerning adjacent coastal States

The Chinese delegation considers that, the rule of "due regard" provided in the UNCLOS is the general standard to deal with the relations between the adjacent coastal States as well as the State conducting activities in the areas beyond national jurisdiction. We suppose that, the new international instrument should address the issue of compatibility between measures under the new international instrument and those established by adjacent coastal States in accordance with the rule of "due regard".

We fully understand the concerns of the relevant adjacent coastal States on this issue. Therefore, we suppose that the new international instrument explicitly provides that, in establishing the measures of ABMTs, including MPAs, the opinions of the adjacent coastal States should be taken into account in appropriate approach.

The main basis for the Chinese delegation to make the above suggestions includes: first, according to the UNCLOS, States shall conduct activities on the high seas or in the Areas with "due regard" for the rights and freedom of other States, including adjacent coastal States. Second, pursuant to the UNCLOS, each State enjoys equal rights in the areas beyond national jurisdiction. The adjacent coastal States do not have any special privileges.

4.2 (c) The manner to respect for the rights of coastal States over all areas under national jurisdiction.

The new international instrument should explicitly provide that, the ABMTs, including MPAs, shall not undermine the rights of coastal States over all areas under their national jurisdiction in accordance with the UNCLOS, including the rights over the exclusive economic zone, and the continental shelf within and beyond 200 nautical miles. Nor shall they undermine the rights of each State in the areas beyond national jurisdiction in accordance with the UNCLOS.

4.3 Process in relation to area – based management tools, including marine protected areas

4.3 (a) – (d) Most appropriate approach for institutional arrangement

The Chinese delegation proposes that, the new international instrument should, without undermining the mandates and functions of existing regional and sectoral bodies, establish the Conference of Parties (CoP) to discharge the function of decision – making and supervision. Under the CoP, a Council and a Secretariat should also be established.

This institutional arrangement focuses on two key points. First, the mandates of existing regional and sectoral bodies, including their existing mandates on the conservation of marine biodiversity, should not be undermined. Second is to establish new institutional arrangement of the CoP to deal with the issue of ABMTs with the objectives of the conservation and sustainable use of marine biodiversity.

The CoP is the highest decision-making body, in charge of reviewing the proposal raised by States Parties for ABMTs, including MPAs, and monitoring. The decision on relevant proposals should be made by States Parties on the basis of consensus.

The Council is the executive body of the CoP, making policies, standards and rules in accordance with the authorization of the new international instrument. Several standing and non-standing committees may be established under the Council, such as scientific and technical committee and legal committee, which are in charge of researching the relevant scientific and legal issues as well as providing advisory opinions. Under the committees, *ad hoc* working groups may be set up when necessary and carry out the work in accordance with the authorization of committees.

The Secretariat, as a permanent institution, is in charge of daily administrative affairs and conduct coordination with other international organizations.

When the measures of ABMTs to be adopted under the new international instrument touch upon the mandates of regional or sectoral bodies in aspect of the conservation of marine biodiversity, consultation and coordination may be conducted between the scientific and technical committee and existing bodies, to explore the most appropriate conservation measures.

4.3.1 Identification of areas

(a) Process

The Chinese delegation considers that, the identification of areas which may need protection should follow the following procedures:

The State which raises a proposal conducts investigation and collects scientific information in accordance with relevant standards. Then it carries out study on the obtained scientific information and makes comparison with the required standards, to identify the scope of protected areas. The final step is to submit the relevant proposal to the CoP for review.

(b) Standard and criteria

The Chinese delegation considers that, the identification of relevant protected areas should be based on the best scientific evidence, and meet both standards and criteria of the bioecological factors and socio-economic factors. The decisions should be made on a case-by-case basis. Among others, with regard to what specific criteria should be included in bioecological factors and socio-economic factors, it is not necessary for the new international instrument to provide an exclusive list. Rather, this should be decided by the competent body under the new international instrument through discussion.

(c) Level of detail

The Chinese delegation suggests that it is advisable to refer to the approach of setting out the standards and criteria of ABMTs adopted in existing international instruments and not to include specific standards and criteria. Instead, relevant standards and criteria should be formulated by the competent body under the framework of the new international instrument.

(d) The possibility of reviewing and/or updating the standards and criteria

The Chinese delegation considers that, the new international instrument may provide that the standards and criteria may be reviewed and updated according to actual needs.

4.3.2 Designation process

(i) Proposal

The Chinese delegation considers that, the proposal to establish a MPA should be raised by States Parties, and be submitted to the CoP which is to be established according to the new international instrument.

The content of the proposal should include the following:

First, general description of the proposed MPA.

Second, description of the objectives and targets of protection.

Third, legal basis, scientific data and factual evidence of the MPA.

Fourth, management plans and measures.

Fifth, scientific research and monitoring plans.

Sixth, duration of protection.

(ii) Consultation on and assessment of the proposal

(a) The scope of participators in the coordination and consultation process should be decided according to the objectives, target and areas of the conservation and the entities concerned. They include but are not limited to States, international organizations, and other stakeholders including non-State entities, such as civil society, industry, scientists and traditional knowledge holders. In order to maintain an open coordination and consultation process, the new international instrument only needs to contain general provisions instead of listing specific stakeholders.

(b) The State which raises a proposal should fully take account of the opinions from States Parties, relevant international organizations and non-State entities. Consultation and coordination should be conducted when necessary.

(c) The new international instrument may establish a scientific and technical committee to evaluate the necessity, scientificity, reasonableness and feasibility of ABMTs. It may also establish a legal committee to provide evaluations and recommendations in legal aspect. The members of the relevant committees are elected from candidates recommended by each State and work in their individual capacity.

(iii) Decision-making

(a) (i) The new international instrument should adopt a consensus-based decision-making model to determine the ABMT related matters.

(a) (ii) The new international instrument may establish the CoP to give effect to the decisions related to ABMTs.

(b) The decisions related to ABMTs should be made on the basis of extensive consultation and comprehensive consideration of the concerns of States Parties and the mandates of relevant competent international organizations, to enhance cooperation and coordination.

(c) Adjacent coastal States may fully participate in the consultation and assessment in the designation process. But in the decision-making procedure, adjacent coastal States have equal status with other States Parties.

4.4 Implementation

The Chinese delegation proposes that, the new international instrument may provide that, States Parties should carry out management measures with respect to activities under their national jurisdiction or control. They should carry out relevant activities according to research and monitoring plans. Meanwhile, international cooperation should be encouraged between States Parties, as well as between States Parties and international organizations, to promote the achievement of objectives of ABMTs.

4.5 Monitoring and review

The Chinese delegation proposes that, the new international instrument should regulate the monitoring and assessment with respect to ABMTs, specifying that the scientific and technical committee subordinated to the CoPs takes charge of assessment and recommendations, which should be submitted to the CoPs for the review. The CoPs will also raise adaptive management measures.

4.6 Issues from the cross-cutting elements

4.6.1 Use of terms

The Chinese delegations proposes that, the new international instrument should provide the definition of the terms "area-based management tools" and "marine protected areas".

4.6.2 Relationship to the Convention and other instruments and frameworks and relevant global, regional and sectoral bodies

The new international instrument should not undermine the relevant legal instruments or frameworks, including the UNCLOS and its two implementing agreements. In this regard, Article 4 of the 1995 Fish Stocks Agreement, which reads "nothing in this Agreement shall prejudice the rights, jurisdiction and duties of States under the Convention", may provide guidance on establishing the relationship between the new international instrument and the UNCLOS. The new international instrument should not undermine the mandates of the existing relevant global, regional and sectoral bodies, including FAO, RFMOs and RFMAs, ISA and IMO. The new international instrument should promote coordination and cooperation with the existing relevant international bodies, to avoid overlaps or conflicts of their mandates.

4.6.3 General principles and approaches

(a) The Chinese delegation considers that, in addition to the general principles applied to the new international instrument, the following principles and approaches should be provided in this regard:

The first principle is that a balanced consideration should be given to interests of coastal States, other States and international community as a whole.

The second is the principle of equal emphasis on the conservation and sustainable use. The conservation and sustainable use of biodiversity are two objectives of ABMTs. A reasonable balance should be struck between the two, rather than favoring one more than another.

The third is the integrated management principle. According to the Preamble of the UNCLOS, "conscious that the problems of ocean space are closely interrelated and need to be considered as a whole".

The fourth is the best available scientific evidence principle. The adoption of ABMTs needs consolidated scientific evidence, to assess potential threats and risks of the protected ecosystem, habitats and species. Without the best available scientific evidence, it is inappropriate to adopt ABMTs.

The fifth is the differential protection principle. Different ABMTs should be adopted in accordance with the respective characteristics of different marine areas, ecosystems, habitats and species.

The sixth is the international cooperation and coordination principle. States and international organizations should enhance cooperation and coordination with respect to the adoption of ABMTs, including MPAs.

The seventh is the mutual due regard principle. With respect to area-based management, mutual due regard should be taken between coastal States and other States, and meanwhile, each Party should have regard to the interests of international community as a whole.

The eighth is the principle of necessity and proportionality. ABMTs is tool, instead of purpose. Real necessity should be the premise of the adoption of ABMTs. Meanwhile, protection measures should be adapt to protection objectives and effects, and be applied on the premise of being cost effective.

(b) The new international instrument may contain general provisions with regard to the above principles. Meanwhile, it may require that documents applied to ABMTs, including management plans and measures, as well as research and monitoring plans, contain explanation or specific provisions to give effect to the above principles.

4.6.4 International cooperation

The new international instrument should, in the light of the spirit of the UNCLOS, provide that States Parties and competent international organizations shall carry out international cooperation with respect to ABMTs including MPAs in accordance with the principle of respect for sovereignty and jurisdiction. Such cooperation should aim at the conservation and sustainable use of marine biodiversity and be based on mutual benefits.

4.6.5 Institutional arrangements

The Chinese delegation proposes to set up the CoPs and its subsidiary bodies as institutional arrangements.

4.6.6 Clearing-house mechanism

The Chinese delegation proposes that, the new international instrument should, in line with the principle of consultation, contribution and shared benefits, establish a public and transparent information exchange platform shared by all stakeholders. Such information platform is designed to integrate resources of various sources, not only through incorporating the existing intergovernmental or non-governmental information exchange platforms with links, such as IOC and OBIS, but also through collecting and storing information and data related to BBNJ, such as that of ABMTs including MPAs. Such information may specifically cover States' proposals with respect to ABMTs, decision of competent authorities, specific management measures and research and monitoring plans, regular review results and subsequent action plans.

5.1 Obligation to conduct environmental impact assessments

Articles 204-206 of the UNCLOS explicitly provide the obligations of States Parties to monitor, assess and report the environmental impacts of activities under their jurisdiction or control. The new international instrument should give effect to the provisions of the UNCLOS and provide a general obligation for each State to assess potential impacts of planned activities under its jurisdiction or control in ABNJ. It should also encourage States Parties by taking domestic legislative, administrative and policy measures, to establish regime and procedures for the conduction of environmental impact assessments (EIAs) prior to planned activities in ABNJ, including the participation of stakeholders and the public.

5.2 Relationship to EIA processes under relevant instruments, frameworks and bodies

The Chinese delegation considers that, the new international instrument should respect the mandates and roles of existing international instruments, frameworks or bodies in EIAs in their respective fields, avoiding establishing new EIAs rules for the same category of activities. It should not undermine the rights and obligations of States under existing international instruments and frameworks.

5.3 Activities for which an EIA is required

First, with respect to the thresholds and criteria for EIAs, the new international instrument should, in accordance with Article 206 of the

UNCLOS, provide the threshold for triggering EIAs is that States "have reasonable grounds for believing…may cause substantial pollution or significant and harmful changes". At present, the existing international instruments do not provide a standard to decide what is "substantial pollution or significant and harmful changes". States may make a comprehensive determination through referring to existing EIAs practices and taking into account of factors such as the characteristics, location and impact features of planned activities as well as the capacity to react to impacts. The new international instrument may also develop corresponding guidelines in this regard.

Second, with regard to whether to develop a list, the Chinese delegation considers that, the impacts on marine environment brought by different categories of maritime activities are not necessarily same. Such impacts not only depend on the type of activities, but also on the scale, intensity, location of activities and how they will impact. The approach of including a list has certain limitations. But if all parties regard a list as necessary, we consider such list should be of open and recommended nature and for reference only, without mandatory binding force.

Third, with regard to cumulative impacts, the Chinese delegation considers that cumulative impacts could be one of factors for consideration when assessing the environmental impacts of activities, rather than a single type of EIAs.

Fourth, with regard to whether to include a specific provision for EIAs in areas identified as ecologically or biologically significant or vulnerable, the Chinese delegation considers that, when deciding whether to trigger EIAs, States have taken into account comprehensively the special location, characteristics and impacts features of relevant activities, including whether the activities are located in areas with significantly environmental sensitivity, vulnerability and representativity. Therefore, it is not necessary for the new international instrument to include a specific provision in this regard.

5.4 EIA process

First, with regard to procedural steps of EIAs, those provided in existing international instruments generally include: screening, public notification and consultation, publication of reports and public availability of reports, consideration of reports, publication of decision-making documents, monitoring and review. The Chinese delegation considers that, the Prepcom Report has covered all the above-mentioned steps. Additional procedures and steps are not necessary.

Second, with regard to the level of detail regarding procedural steps for EIA in the new international instrument, the Chinese delegation proposes that the new international instrument should adopt a general approach to EIA process. Detailed contents are to be specified in forms of recommendations or guidelines.

Third, with regard to which aspects regarding EIA process should be internationalized, the Chinese delegation proposes that, the new international instrument should adopt a State-driven approach to EIAs. The initiation, decision-making and implementation lie on States. The EIA process may be to some extent internationalized at the phases of publication of EIA draft reports and public consultation.

Fourth, with regard to the participation of adjacent coastal States, the new international instrument should not include transboundary impacts of activities within national jurisdiction. Only when activities carried out in areas beyond national jurisdiction may have significant environmental impacts on marine areas within coastal States' jurisdiction, may the participation of adjacent coastal States be concerned. Opinions of adjacent coastal States may be listened to at the initiative phase of EIAs. They may be invited to make comments and recommendations, But decisions should be made by States initiating EIAs.

5.5 Content of EIA reports

(a) With regard to the content of EIA reports, the Chinese delegation proposes that, it may include: description of the planned activities and their purpose; description of reasonable alternatives to the planned activities; description of marine environment and ecosystem which may be significantly impacted by the planned activities and their alternatives; description of the potential effects of the planned activities on the marine environment and ecosystem; description of any measures for avoiding, preventing and mitigating impacts; the development of environmental management and monitoring plans.

(b) With respect to the level of detail of EIA reports, the Chinese delegation proposes that, the new international instrument only lists a general framework for EIA reports. For example, it may be referred to Annex 1 to the Protocol on Environment Protection to the Antarctic Treaty and Annex 2 to the Convention on Environmental Impact Assessment in a Transboundary Context, which only list a general framework for EIA reports.

(c) With respect to whether transboundary impacts should be addressed in an activity - oriented approach or an impact - oriented approach, the Chinese delegation proposes that assessments should be conducted on activities carried out beyond national jurisdiction and may have significant environment impacts on the marine areas within coastal States' jurisdiction.

5.6 Monitoring, reporting and review

The first point is that, according to Articles 204 and 205 of the

UNCLOS and existing international instruments and practices, the new international instrument should provide that the monitoring, reporting and review on EIAs are carried out primarily by States. The main role played by the third parties or international bodies regarding assessment and review is to provide recommendations.

The second point is regarding which information should be provided to those adjacent States potentially affected by transboundary impacts. The initiator of activities should inform the affected side of the information of the planned activities, the EIA procedures and that significant transboundary impacts may be caused by the planned activities. The affected side should determine whether to participate in the EIA procedures and provide the initiator of activities with the information concerning the potential significant transboundary impacts. After the consultation on transboundary EIAs, the initiator provides the affected side with the final decision concerning the planned activities and reasons for such decision.

5. 7 Strategic environmental assessments

According to Article 206 of the UNCLOS, the Chinese side considers that, the target of EIAs in the new international instrument should be the panned "activities" under the jurisdiction or control of States, not including strategic environmental assessments (SEA). We notice that some States propose to carry out SEA. But what is SEA, what to assess and how to carry out SEA are not clear. It also lacks international practices. The Chinese delegation looks forward to further explanation by the relevant States.

5. 8 Issues from the cross-cutting elements

5. 8. 1 Use of terms

The Chinese delegation proposes that, the definition of the term "environmental impact assessment" should be considered to be included in the new international instrument.

5.8.2 Relationship to the Convention and other instruments and frameworks and relevant global, regional and sectoral bodies

The Chinese delegation considers that, the regime regarding EIAs in the new international instrument, which is an implementing agreement of the UNCLOS, should follow the basic legal framework and procedural elements established in the UNCLOS, in particular Article 206 of the UNCLOS. Meanwhile, such regime should also take into account the provisions on EIAs in other international instruments. The new international instrument should not undermine the mandates of existing relevant global, regional and sectoral bodies, including ISA and RFMOs. The new international instrument should promote coordination and cooperation with existing relevant international bodies and avoid the overlap and conflict of their mandates.

5.8.3 General principles and approaches

With respect to EIA matters, the new international instrument should include the following principles and approaches:

The first is the principle of equal emphasis on the conservation and sustainable use. The EIA is a preventive measure to protect and preserve marine environment. The relevant regime should not only have benefit to the promotion of marine environmental protection, but also be in accordance with the objective of sustainable use.

The second is the State-led principle. The initiation, conduction and decision-making of EIAs should all be carried out by States.

The third is the best available scientific evidence principle.

The fourth is the practical principle. EIAs should be technically practical and cost-effective.

The fifth is public and transparency principle.

5. 8. 4 International cooperation

The new international instrument should encourage States to conduct cooperation on EIAs, including the joint EIAs carried out by two or more States.

5. 8. 5 Institutional arrangements

The Chinese delegation suggests, the new international instrument stipulates that each State Party should report to the Conference of Parties (CoPs) about the conduction of EIAs.

5. 8. 6 Clearing-house mechanism

The Chinese delegation considers that, regarding EIAs, the clearing-house mechanism may play a role in the publication of the draft of EIA reports and consultation, the publication of reports on EIA results, the access to States Parties' policies, guidelines and technical methods on EIAs, the promotion of capacity building such as relevant education and training, the exchange of best practices of EIAs and so forth.

6. 1 Objectives of capacity building and the transfer of marine technology

First, the new international instrument may provide that the objective of capacity building and the transfer of technology is to promote the exploration, understanding, conservation and sustainable use of marine biodiversity beyond national jurisdictions.

Second, the new international instrument may provide that each State shall, within its capacity, promote international cooperation with developing countries on capacity-building and the transfer of marine technology, to effectively improve developing countries' capacity for the conservation and sustainable use of BBNJ, in particular by taking into account the special needs of the least developed countries, landlocked developing countries, geographically disadvantaged States and small island developing States, as well as coastal African States.

6.2 Types of and modalities for capacity building and transfer of marine technology

With regard to whether a list were to be included, the Chinese delegation is open to the inclusion of an indicative, non-exhaustive list in the new international instruments. The list may include but is not limited to the following: marine scientific knowledge and information exchange on BBNJ, marine science and technology research and application, marine infrastructure, marine policy, technology standards and rules, personnel training and capacity building programs, technical exchange and cooperation on marine science.

With regard to the modalities for capacity-building and the transfer of technology, the new international instrument should fully draw on the modalities of the existing international organizations, such as the Intergovernmental Oceanographic Commission, to strengthen capacity-building and transfer of marine technology. We could also consider to establish a new mechanism on the basis of extensive consultation to strengthen international cooperation and information sharing on the capacity building and the transfer of technology.

With regard to the terms and conditions on the transfer of technology, the transfer of marine technology should encourage scientific research, promote technological innovation and respect for intellectual property rights, and the terms of technology transfer should be freely arrived at between the supplier and the recipient on the basis of equality and voluntariness, fairness and reasonableness, as well as mutual benefits and reciprocity.

With regard to a clearing-house mechanism, the new international instrument should, on the basis of the principle of consultation, contribution and shared benefits, establish an open and transparent clearing house mechanism shared by all stakeholders. Aiming at integrating resources from all sources, the clearing-house mechanism may not only incorporate the existing intergovernmental and non - governmental information exchange platforms through establishing links, such as the Intergovernmental Oceanographic Commission (IOC) and Ocean Biogeographic Information System (OBIS), but also collect and store the relevant information and data on BBNJ, including those on capacity-building and the transfer of technology.

6.4 Monitoring and Review of capacity building and the transfer of technology

The Chinese delegation suggests that, with respect to capacity building and the transfer of technology, the new instrument should establish a regime requiring the submission of performance reports to the Conference of Parties (CoP). Under such regime, each State Party should report the progress of the relevant work to the CoP. The CoP will review performance reports and make recommendations as guidance, not as mandates.

6.5 Issues from the cross-cutting elements

6.5.1 Use of terms

The Chinese delegation suggests that the definition of the two terms-"capacity building" and "the transfer of marine technology" -may be considered to be included in the instrument.

6.5.2 Relationship to the Convention and other instruments and frameworks and relevant global, regional and sectoral bodies

The Chinese delegation proposes that the new international instrument should not undermine the existing relevant legal instrument or framework, including the UNCLOS and its two implementing agreement. Nor should it interfering with the mandates of the existing relevant global, regional and sectoral bodies, such as the Food and Agriculture Organization of the United Nations (FAO), Regional Fisheries Management Organizations and Arrangements (RFMO/As), the International Seabed Authority (ISA) and the International Maritime Organization (IMO). The new international instrument should facilitate coordination and cooperation with the existing relevant international bodies and avoid overlap or conflict of functions.

6.5.3 General principles and approaches

The Chinese delegation proposes that the new instrument should be based on the regime established in Part XIV of the UNCLOS, following the principles of pertinence, effectiveness, equality and voluntariness, win-win cooperation, the protection of legitimate rights and interests including intellectual property rights in accordance with law, and the preferential treatment to developing countries.

6.5.4 International cooperation

The Chinese delegation proposes that the new international instrument should encourage each State to develop various forms of international cooperation in each field, to continuously improve each State's capacity for and level of the understanding, conservation and sustainable use of BBNJ.

附　　录

2017 年 12 月 24 日大会决议

根据《联合国海洋法公约》的规定就国家管辖范围以外区域海洋生物多样性的养护和可持续利用问题拟订一份具有法律约束力的国际文书

大会，遵循《联合国宪章》所载宗旨和原则，回顾其 2015 年 6 月 19 日第 69/292 号决议，注意到大会第 69/292 号决议所设预备委员会题为"根据《联合国海洋法公约》的规定就国家管辖范围以外区域海洋生物多样性的养护和可持续利用问题拟订一份具有法律约束力的国际文书"的报告。[①]

1. 决定在联合国主持下召开一次政府间会议，审议预备委员会关于案文内容的建议，并为根据《联合国海洋法公约》[②] 的规定就国家管辖范围以外区域海洋生物多样性的养护和可持续利用问题拟订一份具有法律约束力的国际文书拟订案文，以尽早制定该文书；

2. 又决定谈判应处理 2011 年商定的一揽子事项中确定的专

① A/AC. 287/2017/PC. 4/2。

② 联合国，《条约汇编》，第 1833 卷，第 31363 号。

题，即国家管辖范围以外区域海洋生物多样性的养护和可持续利用，特别是作为一个整体的全部海洋遗传资源的养护和可持续利用，包括惠益分享问题，以及包括海洋保护区在内的划区管理工具、环境影响评估和能力建设及海洋技术转让等措施；

3. 又决定，最初在 2018 年、2019 年和 2020 年上半年召开四届会议，每次会期为 10 个工作日，第一届会议在 2018 年下半年举行，第二和第三届会议将于 2019 年举行，第四届会议将在 2020 年上半年举行，并请秘书长在 2018 年 9 月 4 日至 17 日召开第一届会议；

4. 决定会议应于 2018 年 4 月 16 日至 18 日在纽约举行为期三天的组织会议，讨论组织事项，包括文书预稿的起草过程；

5. 请大会主席以公开透明方式，就会议候任主席或候任共同主席的提名进行磋商；

6. 重申会议的工作和成果应完全符合《联合国海洋法公约》的规定；

7. 认识到这一进程及其结果不应损害现有有关法律文书和框架以及相关的全球、区域和部门机构；

8. 决定会议应向联合国所有会员国、专门机构成员和《公约》缔约方开放；

9. 强调指出必须确保尽可能广泛和有效地参加会议；

10. 认识到参加谈判和谈判结果都不可影响《公约》或任何其他相关协议的非缔约国在涉及这些文书方面的法律地位，也不可影响《公约》或任何其他相关协议的缔约国在涉及这些文书方面的法律地位；

11. 决定，就该会议的各次会议而言，已加入《公约》的国际组织的参与权应与《公约》缔约国会议的参与权相同，本规定对所有适用大会 2011 年 5 月 3 日第 65/276 号决议的会议不构成先例；

12. 又决定邀请已收到大会根据其有关决议发出的长期邀请的组织和其他实体的代表，以观察员身份参加其会议和工作，但前提是这些代表将以这一身份参加会议，并邀请获邀参加相关主要会议和首脑会议的有关全球和区域政府间组织及其他有关国际机构的代表以会议观察员身份参加会议①；

13. 还决定按照经济及社会理事会1996年7月25日第1996/31号决议的规定，会议也向具有经济及社会理事会咨商地位的有关非政府组织并向已获得认可参加各次主要会议和首脑会议的有关非政府组织开放②，它们可作为观察员出席会议，但有一项谅解，即除非会议在具体情况下另有决定，参与意味着出席正式会议，获得正式文件副本，将它们的材料提供给代表，及酌情让它们当中数量有限的代表在会上发言；

14. 决定邀请区域委员会准成员③以观察员身份参与会议的工作；

15. 又决定邀请联合国系统有关专门机构以及其他机构、组织、基金和方案的代表作为观察员出席；

---

① 应邀参加下列有关主要会议和首脑会议的政府间组织和其他国际机构：可持续发展问题世界首脑会议；联合国可持续发展大会和之前在巴巴多斯、毛里求斯和萨摩亚举行的小岛屿发展中国家可持续发展问题联合国会议；联合国跨界鱼类种群和高度洄游鱼类种群会议；执行1982年12月10日联合国海洋法公约有关养护和管理跨界鱼类种群和高度洄游鱼类种群的规定的协定审查会议；联合国支持落实可持续发展目标14即保护和可持续利用海洋和海洋资源以促进可持续发展会议。

② 已获得认可参加下列有关主要会议和首脑会议的非政府组织：可持续发展问题世界首脑会议；联合国可持续发展大会和之前在巴巴多斯、毛里求斯和萨摩亚举行的小岛屿发展中国家可持续发展问题联合国会议；联合国支持落实可持续发展目标14即保护和可持续利用海洋和海洋资源以促进可持续发展会议。

③ 美属萨摩亚、安圭拉、阿鲁巴、百慕大、英属维尔京群岛、开曼群岛、北马里亚纳群岛联邦、库拉索岛、法属波利尼西亚、关岛、蒙特塞拉特、新喀里多尼亚、波多黎各、圣马丁岛、特克斯和凯科斯群岛及美属维尔京群岛。

16. 还决定向会议转递预备委员会的报告；

17. 决定会议应秉诚并尽一切努力，以协商一致方式商定实质性事项；

18. 又决定，除本决议第 17 和 19 段的规定外，除非会议另有协议，有关大会程序和惯例的规则应适用于该会议的程序；

19. 还决定，在不违反第 17 段的情况下，会议关于实质性事项的决定应以出席并参加表决的代表的三分之二多数作出，在此之前，主持人应通知会议，已为通过协商一致方式达成协议竭尽一切努力；

20. 回顾大会邀请会员国、国际金融机构、捐助机构、政府间组织、非政府组织和自然人和法人向第 69/292 号决议所设自愿信托基金提供捐款，并授权秘书长扩大该信托基金提供的援助，以便除支付经济舱旅费外还包括每日生活津贴，但每届会议向该信托基金提出的援助申请仅限每个国家一位代表；

21. 请秘书长任命一位会议秘书长作为秘书处内部协调人，为会议组织工作提供支持；

22. 又请秘书长向会议提供开展工作所需的协助，包括提供秘书处服务和必要的背景资料和相关文件，并安排由秘书处法律事务厅海洋事务和海洋法司提供支持；

23. 决定继续处理此案。

2017 年 12 月 24 日
第 76 次全体会议

Resolution adopted by the General Assembly on 24 December 2017

International legally binding instrument under the United Nations Convention on the Law of the Sea on the conservation and sustainable use of marine biological diversity of areas beyond national jurisdiction

The General Assembly, Guided by the purposes and principles enshrined in the Charter of the United Nations, Recalling its resolution 69/292 of 19 June 2015, Taking note of the report of the Preparatory Committee established by General Assembly resolution 69/292: Development of an international legally binding instrument under the United Nations Convention on the Law of the Sea on the conservation and sustainable use of marine biological diversity of areas beyond national jurisdiction,① and the recommendations contained therein。

1. Decides to convene an intergovernmental conference, under the auspices of the United Nations, to consider the recommendations of the Preparatory Committee on the elements and to elaborate the text of an international legally binding instrument under the United Nations Convention on the Law of the Sea② on the conservation and sustainable use of marine bio-

① A/AC. 287/2017/PC. 4/2.

② United Nations, Treaty Series, vol. 1833, No. 31363.

logical diversity of areas beyond national jurisdiction, with a view to developing the instrument as soon as possible;

2. Alsodecides that negotiations shall address the topics identified in the package agreed in 2011, namely, the conservation and sustainable use of marine biological diversity of areas beyond national jurisdiction, in particular, together and as a whole, marine genetic resources, including questions on the sharing of benefits, measures such as area-based management tools, including marine protected areas, environmental impact assessments and capacity-building and the transfer of marine technology;

3. Further decides that, initially with respect to 2018, 2019 and the first half of 2020, the conference shall meet for four sessions of a duration of 10 working days each, with the first session taking place in the second half of 2018, the second and third sessions taking place in 2019, and the fourth session taking place in the first half of 2020, and requests the Secretary-General to convene the first session of the conference from 4 to 17 September 2018;

4. Decides that the conference shall hold a three-day organizational meeting in New York, from 16 to 18 April 2018, to discuss organizational matters, including the process for the preparation of the zero draft of the instrument;

5. Requests the President of the General Assembly to undertake consultations, in an open and transparent manner, for the nomination of a President-designate or co- Presidents-designate of the conference;

6. Reaffirms that the work and results of the conference should be fully consistent with the provisions of the United Nations Convention on the Law of the Sea;

7. Recognizes that this process and its result should not undermine

existing relevant legal instruments and frameworks and relevant global, regional and sectoral bodies;

8. Decides that the conference shall be open to all States Members of the United Nations, members of the specialized agencies and parties to the Convention;

9. Stresses the need to ensure the widest possible and effective participation in the conference;

10. Recognizes that neither participation in the negotiations nor their outcome may affect the legal status of non-parties to the Convention or any other related agreements with regard to those instruments, or the legal status of parties to the Convention or any other related agreements with regard to those instruments;

11. Decides that, for the meetings of the conference, the participation rights of the international organization that is a party to the Convention shall be as in the Meeting of States Parties to the Convention and that this provision shall constitute no precedent for all meetings to which General Assembly resolution 65/276 of 3 May 2011 is applicable;

12. Alsodecides to invite to the conference representatives of organizations and other entities that have received a standing invitation from the General Assembly pursuant to its relevant resolutions to participate, in the capacity of observer, in its sessions and work, on the understanding that such representatives would participate in the conference in that capacity, and to invite, as observers to the conference, representatives of interested global and regional intergovernmental organizations and other interested international bodies that were invited to participate

in relevant conferences and summits[①];

13. Further decides that attendance at the conference as observers will also be opened to relevant non-governmental organizations in consultative status with the Economic and Social Council in accordance with the provisions of Council resolution 1996/31 of 25 July 1996, as well as to those that were accredited to relevant conferences and summits[②], on the understanding that participation means attending formal meetings, unless otherwise decided by the conference in specific situations, receiving copies of the official documents, making available their materials to delegates and addressing the meetings, through a limited number of their representatives, as appropriate;

---

① Reference is made to intergovernmental organizations and other international bodies that were invited to participate in the following relevant conferences and summits: The World Summit on Sustainable Development, the United Nations Conference on Sustainable Development and the previous United Nations conferences on sustainable development of small island developing States, held in Barbados, Mauritius and Samoa, the United Nations Conference on Straddling Fish Stocks and Highly Migratory Fish Stocks, the Review Conference on the Agreement for the Implementation of the Provisions of the United Nations Convention on the Law of the Sea of 10 December 1982 relating to the Conservation and Management of Straddling Fish Stocks and Highly Migratory Fish Stocks, as well as the United Nations Conference to Support the Implementation of Sustainable Development Goal 14: Conserve and sustainably use the oceans, seas and marine resources for sustainable development.

② Reference is made to the non-governmental organizations that were accredited to the following relevant conferencesand summits: The World Summit on Sustainable Development, the United Nations Conference on Sustainable Development and the previous United Nations conferences on sustainable development of small island developing States, held in Barbados, Mauritius and Samoa, as well as the United Nations Conference to Support the Implementation of Sustainable Development Goal 14: Conserve and sustainably use the oceans, seas and marine resources for sustainable development.

14. Decides to invite associate members of regional commissions [①] to participate in the work of the conference in the capacity of observer;

15. Alsodecides to invite representatives of relevant specialized agencies, as well as other organs, organizations, funds and programmes of the United Nations system as observers;

16. Further decides to forward the report of the Preparatory Committee to the conference;

17. Decides that the conference shall exhaust every effort in good faith to reach agreement on substantive matters by consensus;

18. Alsodecides that, except as provided for in paragraphs 17 and 19 of the present resolution, the rules relating to the procedure and the established practice of the General Assembly shall apply to the procedure of the conference unless otherwise agreed by the conference;

19. Further decides that, subject to paragraph 17, decisions of the conference on substantive matters shall be taken by a two-thirds majority of the representatives present and voting, before which, the presiding officer shall inform the conference that every effort to reach agreement by consensus has been exhausted;

20. Recalls its invitation to Member States, international financial institutions, donor agencies, intergovernmental organizations, non-governmental organizations and natural and juridical persons to make financial contributions to the voluntary trust fund established in resolution69/292, and authorizes the Secretary-General to expand the assis-

---

① American Samoa, Anguilla, Aruba, Bermuda, the British Virgin Islands, the Cayman Islands, the Commonwealth of the Northern Mariana Islands, Curaçao, French Polynesia, Guam, Montserrat, New Caledonia, Puerto Rico, Sint Maarten, the Turks and Caicos Islands and the United States Virgin Islands.

tance provided by this trust fund to include daily subsistence allowance in addition to defraying the costs of economy-class travel, limiting requests for assistance from this trust fund to one delegate per State for each session;

21. Requests the Secretary-General to appoint a Secretary-General of the conference to serve as focal point within the Secretariat for providing support to the organization of the conference;

22. Also requests the Secretary-General to provide the conference with the necessary assistance for the performance of its work, including secretariat services and the provision of essential background information and relevant docume nts, and to arrange for support to be provided by the Division for Ocean Affairs and the Law of the Sea of the Office of Legal Affairs of the Secretariat;

23. Decides to remain seized of the matter.

76th plenary meeting 24 December 2017

大会关于根据《联合国海洋法公约》的规定就国家管辖范围以外区域海洋生物多样性的养护和可持续利用问题拟订一份具有法律约束力的国际文书的第 69/292 号决议所设预备委员会的报告

一、导言

1. 2015 年 6 月 19 日大会第 69/292 号决议决定根据《联合国海洋法公约》(《公约》)的规定就国家管辖范围以外区域海洋生物多样性的养护和可持续利用问题拟订一份具有法律约束力的国际文书。为此，大会决定在举行政府间会议之前，设立一个预备委员会，所有联合国会员国、专门机构成员和《公约》缔约方均可参加，并按照联合国惯例邀请其他方面作为观察员参加，以便考虑到共同主席有关不限成员名额非正式特设工作组研究国家管辖范围以外区域海洋生物多样性的养护和可持续利用问题相关工作的各种报告，就根据《公约》的规定拟订一份具有法律约束力的国际文书的案文草案要点向大会提出实质性建议。[①]

2. 大会还决定预备委员会将在 2016 年开始工作，并在 2017

① 见 A/61/65、A/63/79 和 Corr. 1、A/65/68、A/66/119、A/67/95、A/68/399、A/69/82、A/69/177 和 A/69/780。

年底之前向大会报告进展情况，大会将在其第七十二届会议结束之前，考虑到预备委员会的上述报告，就在联合国主持下召开一次政府间会议以及会议的开始日期作出决定，会议目的是审议预备委员会关于要点的建议并根据《公约》的规定拟订具有法律约束力的国际文书案文。

3. 大会确认任何根据《公约》的规定就国家管辖范围以外区域海洋生物多样性问题拟订的具有法律约束力的文书都应确保得到尽可能广泛的接受，并为此决定预备委员会应竭尽一切努力，以协商一致方式就实质性事项达成协议。大会又确认，至关重要的是预备委员会要以有效方式开展工作，根据《公约》的规定拟订一份具有法律约束力的国际文书的案文草案要点，还确认即使在竭尽一切努力后仍未就一些要点达成协商一致，也可将这些要点列入预备委员会向大会提交建议的某一章节之中。

4. 大会决定通过谈判处理 2011 年商定的一揽子事项所含的专题（见第 66/231 号决议），即国家管辖范围以外区域海洋生物多样性的养护和可持续利用问题，特别是共同且作为一个整体处理海洋遗传资源包括惠益分享问题、划区管理工具包括海洋保护区等措施、环境影响评估以及能力建设和海洋技术转让。

5. 大会还确认该进程不应损害现有相关法律文书和框架以及相关的全球、区域和部门机构，参加谈判和谈判结果都不可影响《公约》或任何其他相关协议的非缔约国对于这些文书的法律地位，也不可影响《公约》或任何其他相关协议的缔约国对于这些文书的法律地位。

6. 根据第 69/292 号决议第 6 段，秘书处法律事务厅海洋事务和海洋法司向预备委员会提供了实务秘书处支助。

二、组织事项

A. 预备委员会届会

7. 大会在第 69/292 号决议中决定，预备委员会应在 2016 年和 2017 年举行至少两次会议，每次为期 10 个工作日，根据该决议，秘书长分别于 2016 年 3 月 28 日至 4 月 8 日和 8 月 26 日至 9 月 9 日在联合国总部召开了预备委员会第一届和第二届会议。根据第 71/257 号决议，秘书长分别于 2017 年 3 月 27 日至 4 月 7 日和 7 月 10 日至 21 日，在联合国总部召开了预备委员会第三届会议和第四届会议。

B. 选举主席团成员

8. 大会第六十九届会议主席萨姆·卡汉巴·库泰萨在 2015 年 9 月 4 日给会员国的信中，依照第 69/292 号决议第 1（d）段的规定，任命特立尼达和多巴哥共和国副常驻代表兼该国常驻联合国代表团临时代办伊登·查尔斯担任预备委员会主席。

9. 大会第 69/292 号决议第 1（e）段决定，预备委员会应选举设立一个主席团，由每个区域组两名成员组成，这 10 名成员应就程序事项协助主席开展一般性工作。依照上述规定，预备委员会在第一届会议上选出了由以下成员组成的主席团：Mohammed Atlassi（摩洛哥）、Thembile Elphus Joyini（南非）、马新民（中国）、Kaitaro Nonomura（日本）、Konrad Marciniak（波兰）、Maxim V. Musikhin（俄罗斯联邦）、Javier Gorostegui Obanoz（智利）、Gina Guillén-Grillo（哥斯达黎加）、Antoine Misonne（比利时）和 Giles Norman（加拿大）。

10. 在其第二届会议上，预备委员会选出 Jun Hasebe（日本）

和 Catherine Boucher（加拿大）为主席团成员，取代已从主席团成员职位上辞任的 Kaitaro Nonomura 和 Giles Norman。鉴于亚洲—太平洋集团达成了分享主席团成员职位的协议，预备委员会还选出 Margo Deiye（瑙鲁）从 2016 年 10 月 28 日起担任主席团的成员。

11. 大会第七十一届会议主席彼得·汤姆森在 2017 年 1 月 24 日致函通知各成员国，伊登·查尔斯表示他不能再担任预备委员会主席。他还指出，在与会员国协商后，他已根据第 69/292 号决议第 1（d）段的规定，指定巴西常驻联合国副代表卡洛斯·塞尔吉奥·索布拉尔·杜阿尔特先生担任预备委员会主席。

12. 在第三届会议上，根据第 69/292 号决议第 1（e）段的规定，考虑到拉丁美洲和加勒比国家集团已经达成的协议，预备委员会选举 Pablo Adrían Arrocha Olabuenaga（墨西哥）和 José Luis Fernandez Valoni（阿根廷）担任主席团成员，取代 Javier Gorostegui Obanoz 和 Gina Guillén-Grillo。考虑到马来西亚作为亚洲—太平洋集团主席提供的资料，根据该集团达成的协议，预备委员会还选出 Jun Hasebe（日本）从 2017 年 5 月 28 日起担任主席团成员，取代 2017 年 5 月 27 日从主席团辞任的马新民。

C. 文件

13. 大会第 69/292 号决议确认在文件方面，预备委员会所有文件，除其议程、工作方案和报告外，都将作为非正式工作文件。预备委员会各届会议的正式文件清单附于本报告之后。

14. 此外，为了协助各项进程，主席依其职责编写了若干非正式文件（见第 21、26 和 32 段），包括主席关于第一、二、三届会议的概述以及关于国家管辖范围以外区域海洋生物多样性的养护和可持续利用问题具有法律约束力的国际文书案文草案要点的精简非正式文件。[1]

15. 应主席邀请，各代表团还就案文草案要点提出了意见，可在海洋事务和海洋法司网站上查阅这些意见。

D. 预备委员会届会程序

16. 大会第69/292号决议第1（i）段确认，至关重要的是预备委员会要以有效方式开展工作，根据《公约》的规定拟订一份具有法律约束力的国际文书的案文草案要点，还确认即使在竭尽一切努力后仍未就一些要点达成协商一致，也可将这些要点列入预备委员会向大会提交建议的某一章节之中。大会在该决议中决定，除了上述第1（i）段的规定外，大会各委员会议事程序的规则和惯例适用于预备委员会的议事程序，对于预备委员会会议而言，作为《公约》缔约方的国际组织的参与权应等同于其对《公约》缔约国会议的参与权，并且，该规定对所有适用大会2011年5月3日第65/276号决议的会议不构成先例。

预备委员会第一届会议

17. 在2016年3月28日预备委员会第1次会议上，主管法律事务副秘书长兼联合国法律顾问作了发言。预备委员会通过了A/AC. 287/2016/PC. 1/L. 1号文件所载届会议程，并同意按照A/AC. 287/2016/PC. 1/L. 2号文件所载暂定工作方案开展工作。

18. 预备委员会第一届会议举行了15次全体会议。来自99个联合国会员国、2个非会员国、联合国5个方案、基金和办事处、联合国系统4个专门机构和有关组织、8个政府间组织和17个非政府组织的代表出席了会议。

19. 预备委员会在其全体会议上听取了一般性发言并审议了下列问题：一项具有法律约束力的国际文书的范围及其与其他文书

---

① 可查阅：www. un. org/depts/los/biodiversity/prepcom. htm。

的关系；具有法律约束力的国际文书的指导方针和原则；海洋遗传资源，包括惠益分享问题；划区管理工具包括海洋保护区等措施；环境影响评估；能力建设和海洋技术转让问题。全体会议还讨论并核准了第二届会议的路线图。

20. 还召开了非正式工作组会议，由以下各位主持：卡洛斯·杜阿尔特（巴西）主持海洋遗传资源包括惠益分享问题非正式工作组；John Adank（新西兰）主持划区管理工具包括海洋保护区等措施非正式工作组；René Lefeber（荷兰）主持环境影响评估非正式工作组；Rena Lee（新加坡）主持能力建设和海洋技术转让问题非正式工作组。

21. 第一届会议之后，根据全体会议讨论并核准的路线图，主席编写了一份届会概况。主席还编写了关于事项和问题群组的指示性建议，以协助各非正式工作组在预备委员会第二届会议上进一步讨论。

预备委员会第二届会议

22. 在 2016 年 8 月 26 日预备委员会第 16 次会议上，主管法律事务助理秘书长作了发言。预备委员会通过了 A/AC. 287/2016/PC. 2/L. 1 号文件所载议程，并同意按照 A/AC. 287/2016/PC. 2/L. 2 号文件所载暂定工作方案开展工作。

23. 预备委员会第二届会议举行了 13 次全体会议。来自 116 个联合国会员国、3 个非会员国、联合国 6 个方案、基金和办事处、联合国系统 5 个专门机构和有关组织、9 个政府间组织和 22 个非政府组织的代表出席了会议。

24. 预备委员会在其全体会议上审议了下列问题：海洋遗传资源，包括惠益分享问题；划区管理工具包括海洋保护区等措施；环境影响评估；能力建设和海洋技术转让；贯穿各领域的问题。全体会议还讨论并核准了第三届会议的路线图。

25. 还召开了非正式工作组会议，由以下各位主持：伊登·查尔斯（特立尼达和多巴哥）[①] 主持海洋遗传资源包括惠益分享问题非正式工作组；John Adank（新西兰）主持划区管理工具包括海洋保护区等措施非正式工作组；René Lefeber（荷兰）主持环境影响评估非正式工作组；Rena Lee（新加坡）主持能力建设和海洋技术转让问题非正式工作组。预备委员会主席伊登·查尔斯（特立尼达和多巴哥）主持贯穿各领域的问题非正式工作组。

26. 第二届会议之后，根据全体会议讨论并核准的路线图，主席编写了一份届会概况。主席还就根据《联合国海洋法公约》的规定拟订一份具有法律约束力的国际文书案文草案要点编写了主席的非正式文件和该文件的补充。

预备委员会第三届会议

27. 在 2017 年 3 月 27 日预备委员会第 29 次会议上，主管法律事务副秘书长兼联合国法律顾问作了发言。预备委员会通过了 A/AC. 287/2017/PC. 3/L. 1 号文件所载议程，并同意按照 A/AC. 287/2017/PC. 3/L. 2 号文件所载暂定工作方案开展工作。

28. 预备委员会第三届会议举行了 9 次全体会议。来自 147 个联合国会员国、2 个非会员国、联合国 5 个方案、基金和办事处、联合国系统 4 个专门机构和有关组织、14 个政府间组织和 19 个非政府组织的代表出席了会议。

29. 海洋环境状况包括社会经济方面问题全球报告和评估经常程序特设全体工作组在项目 7“其他事项”下介绍了国家管辖范围以外区域海洋生物多样性的养护和可持续利用问题第一次全球综合海洋评估的技术摘要未经编辑的预发案文。大会第七十一届会议主席彼得·汤姆森还在该项目下向预备委员会致辞。

30. 预备委员会在其全体会议上审议了下列问题：海洋遗传资源，包括惠益分享问题；划区管理工具包括海洋保护区等措施；

① 鉴于卡洛斯·杜阿尔特先生不能主持，由主席主持该非正式工作组。

环境影响评估；能力建设和海洋技术转让；贯穿各领域的问题。全体会议还讨论并核准了第四届会议的路线图。

31. 还召开了非正式工作组会议，由以下各位主持：Janine Elizabeth Coye-Felson（伯利兹）① 主持海洋遗传资源包括惠益分享问题非正式工作组；Alice Revell（新西兰）主持划区管理工具包括海洋保护区等措施非正式工作组②；René Lefeber（荷兰）主持环境影响评估非正式工作组；Rena Lee（新加坡）主持能力建设和海洋技术转让问题非正式工作组；预备委员会主席卡洛斯·杜阿尔特（巴西）主持贯穿各领域的问题非正式工作组。

32. 第三届会议之后，根据全体会议讨论并核准的路线图，主席编写了一份届会概况。主席还编写了指示性建议，以协助预备委员会就根据《联合国海洋法公约》拟订一份具有法律约束力的国际文书案文草案要点编写向大会提出的建议以及关于该案文草案要点的精简非正式文件。

预备委员会第四届会议

33. 在 2017 年 7 月 10 日预备委员会第 38 次会议上，主管法律事务副秘书长兼联合国法律顾问作了发言。预备委员会通过了 A/AC. 287/2017/PC. 4/L. 1 号文件所载临时议程，并同意按照 A/AC. 287/2017/PC. 4/L. 2 号文件所载暂定工作方案开展工作。

34. 预备委员会第四届会议举行了 10 次全体会议。来自 131 个联合国会员国、2 个非会员国、联合国 2 个方案、基金和办事处、联合国系统 9 个专门机构和有关组织、10 个政府间组织和 23 个非政府组织的代表出席了会议。

35. 预备委员会在其全体会议上听取了一般性发言并审议了就根据《联合国海洋法公约》的规定拟订一份具有法律约束力的国

① 取代卡洛斯·杜阿尔特（巴西），因为杜阿尔特先生成为新任预备委员会主席。

② 取代 John Adank（新西兰），他已通知主席不再担任主持人职务。

际文书的案文草案要点编写实质性建议的问题（见下文第 38 段）。全体会议还审议了预备委员会的报告（见下文第 40 段）。

36. 第一周还召开了非正式工作组会议，由以下各位主持：Janine Elizabeth Coye-Felson（伯利兹）主持海洋遗传资源包括惠益分享问题非正式工作组；Alice Revell（新西兰）主持划区管理工具包括海洋保护区等措施非正式工作组；RenéLefeber（荷兰）主持环境影响评估非正式工作组；Rena Lee（新加坡）主持能力建设和海洋技术转让问题非正式工作组；预备委员会主席卡洛斯·杜阿尔特（巴西）主持贯穿各领域的问题非正式工作组。

37. 在第二周的全体会议期间，许多代表团提议在 2018 年召开一次政府间会议，并将其纳入向大会提出的实质性建议。一些代表团还提议，会议应在 2018 年和 2019 年期间至少举行四轮谈判，每轮为期两周，提供全套会议服务。一些代表团建议，会议应比照适用大会议事规则。其他代表团强调指出，是否召开一次政府间会议以及会议的时间和方式问题应留给大会决定，预备委员会的实质性建议不应包含这方面的任何提议，以避免预先限定大会的讨论。一个代表团认为，在举行政府间会议之前，预备委员会可能需要召开更多届会。

三、预备委员会的建议

38. 在 2017 年 7 月 21 日第 47 次会议上，预备委员会以协商一致方式通过了以下建议。

预备委员会按照大会 2015 年 6 月 19 日第 69/292 号决议举行会议，建议大会：

(a) 审议下文 A 节和 B 节所载要点，以期根据《联合国海洋法公约》的规定就国家管辖范围以外区域海洋生物多样性的养护和可持续利用问题拟订一份具有法律约束力的国际文书。A 节和 B

节的内容并非已形成的共识。A 节包含多数代表团意见一致的非排他性要点。B 节重点突出存在意见分歧的一些主要问题。

A 节和 B 节仅供参考之用，因为它们并不反映讨论过的所有选项。这两节均不妨碍各国在谈判中的立场；

（b）大会应尽快作出决定，是否在联合国主持下召开一次政府间会议，以审议预备委员会关于要点的建议并根据《公约》的规定拟订具有法律约束力的国际文书案文。

A 节

一、序言要点

案文将阐明广泛的背景事项，例如：

- 说明拟订该文书所出于的各种考虑因素，包括主要关切和问题
- 确认在国家管辖范围以外区域海洋生物多样性的养护和可持续利用方面《公约》发挥的核心作用以及现行其他相关法律文书和框架以及相关全球、区域和部门机构的作用
- 确认需要增进合作和协调，以促进国家管辖范围以外区域海洋生物多样性的养护和可持续利用
- 确认需要提供援助，使发展中国家，特别是处于不利地理位置的国家、最不发达国家、内陆发展中国家和小岛屿发展中国家以及非洲沿海国能够有效参与国家管辖范围以外区域海洋生物多样性的养护和可持续利用
- 确认需要一个全面的全球制度，以更好地处理国家管辖范围以外区域海洋生物多样性的养护和可持续利用问题
- 表示坚信一项执行《公约》有关规定的协议最符合这些目的，并有助于维护国际和平与安全
- 申明《公约》、其执行协议或本文书未予规定的问题，仍由一般国际法规则和原则加以规范。

二、一般性要点

1. 用语[①]

案文将提供关键用语的定义，同时注意需要与《公约》及其他相关法律文书和框架中的用语定义保持一致。

2. 适用范围

2.1　地理范围

案文将说明，本文书适用于国家管辖范围以外的区域。

案文将指出，应尊重沿海国对其国家管辖范围内的所有区域，包括对 200 海里以内和以外的大陆架和专属经济区的权利和管辖权。

2.2　属事范围

案文将处理国家管辖范围以外区域海洋生物多样性的养护和可持续利用问题，特别是一并作为一个整体处理海洋遗传资源包括惠益分享问题、划区管理工具包括海洋保护区等措施、环境影响评估以及能力建设和海洋技术转让问题。

案文可以规定不在本文书适用范围内的除外事项，并在处理主权豁免相关问题上与《公约》保持一致。

3. 目标

案文将规定，本文书的目的是通过有效执行《公约》，确保国家管辖范围以外区域海洋生物多样性的养护和可持续利用。

如经商定，案文还可以规定其他目标，例如推进国际合作与协调，以确保实现养护和可持续利用国家管辖范围以外区域海洋生物多样性的总体目标。

4. 与《公约》以及其他文书、框架和相关全球、区域和部门机构的关系

关于与《公约》的关系，案文将指出，文书中的任何内容都

① 一些仅与本文书一个部分有关的具体定义可能列于相关部分中。

不应妨害《公约》规定的各国的权利、管辖权和义务。案文将进一步指出，本文书应参照《公约》的内容并以符合《公约》的方式予以解释和适用。

案文将指出，本文书将促进与现有相关法律文书和框架以及相关全球、区域和部门机构的协调一致性，并对其作出补充。案文还将指出，该文书的解释和适用不应损害现有的文书、框架和机构。

案文可确认，《公约》或任何其他相关协定的非缔约方相对于这些文书的法律地位不受影响。

三、国家管辖范围以外区域海洋生物多样性的养护和可持续利用

1. 一般原则和方法[①]

案文将规定国家管辖范围以外区域海洋生物多样性的养护和可持续利用的一般原则和指导方法。

可能的一般原则和方法包括：

- 尊重《公约》所载之权利、义务和利益的平衡
- 兼顾《公约》有关条款所适当顾及的事项
- 尊重沿海国对其国家管辖范围内所有区域，包括对 200 海里以内和以外的大陆架和专属经济区的权利和管辖权
- 尊重各国主权和领土完整
- 只为和平目的利用国家管辖范围以外区域的海洋生物多样性
- 促进国家管辖范围以外区域海洋生物多样性的养护和可持续利用两方面
- 可持续发展
- 在所有各级开展国际合作与协调，包括南北、南南和三方

① 其中一些原则和方法将列于一个单独条款中，有些则列在序言部分。

合作

• 相关利益攸关方的参与

• 生态系统方法

• 风险预防办法

• 统筹办法

• 基于科学的办法，利用现有的最佳科学资料和知识，包括传统知识

• 适应性管理

• 建设应对气候变化影响的能力

• 符合《公约》不将一种污染转变成另一种污染的义务

• “谁污染谁付费”原则

• 公众参与

• 透明度和信息的可取得性

• 小岛屿发展中国家和最不发达国家的特别需要，包括避免直接或间接地将过度的养护行动负担转嫁给发展中国家

• 诚信

2. 国际合作

案文将规定各国有义务合作，以养护和可持续利用国家管辖范围以外区域的海洋生物多样性，并将详细规定这种义务的内容和方式。

3. 海洋遗传资源，包括惠益分享问题

3.1　范围[①]

案文将规定本文书这个章节在地域和属事方面的适用范围。

3.2　获取和惠益分享

3.2.1　获取

案文将述及获取问题。

① 也可以在文书开头关于范围的总章论述范围问题（例如，见上文第二.2部分）

3.2.2　惠益分享

（一）目标

案文将规定，惠益分享的目标是：

• 促进国家管辖范围以外区域海洋生物多样性的养护和可持续利用。

• 建设发展中国家获取和利用国家管辖范围以外区域海洋遗传资源的能力。案文还可列明商定的其他目标。

（二）惠益分享的指导原则和方法[①]

案文将规定惠益分享的指导原则和方法，例如：

• 惠及当代后世

• 促进海洋科学研究以及研究和开发

（三）惠益

案文将规定可以分享的惠益类型。

（四）惠益分享模式

案文将规定惠益分享模式，同时考虑到现有的文书和框架，例如它可以作出安排，建立一个有关惠益分享的信息交换机制。[②]

3.2.3　知识产权

案文可规定本文书与知识产权之间的关系。

3.3　监测国家管辖范围以外区域海洋遗传资源的利用

案文将处理监测国家管辖范围以外区域海洋遗传资源的利用。

4. 划区管理工具包括海洋保护区等措施

4.1　划区管理工具包括海洋保护区的目标

案文将规定划区管理工具包括海洋保护区在养护和可持续利

① 也可以在文书开头关于原则和方法的总章内载列各项原则（例如，见上文第三.1部分）。

② 信息交换机制的功能可以载于文书的单独章节，专门讨论信息交换机制（例如，见下文第五部分），或载于本节。

用国家管辖范围以外区域海洋生物多样性方面的目标。

4.2　与相关文书、框架和机构所规定措施的关系

案文将规定本文书项下措施与现有的相关法律文书和框架以及相关全球、区域和部门机构所定措施之间的关系，目的是促成各种努力之间的一致性和协调性。

案文将申明，就划区管理工具包括海洋保护区而言，必须加强相关法律文书和框架以及相关全球、区域和部门机构之间的合作与协调，不妨碍其各自任务。

案文还将处理该文书规定的措施与毗邻沿海国所制定措施之间的关系问题，包括兼容性问题，不得妨碍沿海国的权利。

4.3　划区管理工具包括海洋保护区的有关程序

考虑到各种类型的划区管理工具，包括海洋保护区，案文会根据将要拟订的方法，规定划区管理工具包括海洋保护区的相关程序以及有关作用和职责。

4.3.1　确定区域

案文将规定，需要保护区域的确定程序将以现有的最佳科学资料、标准和准则为依据，包括：

- 独特性
- 稀有性
- 对物种的生命史各阶段特别重要
- 对受威胁物种、濒危物种或数量不断减少的物种和（或）生境的重要性
- 脆弱性
- 脆性
- 敏感性
- 生物生产力
- 生物多样性
- 代表性

- 依赖性
- 自然度
- 连通性
- 生态过程
- 经济和社会因素。

4.3.2　指定程序

（一）提案

案文将包含关于划区管理工具包括海洋保护区有关提案的条款。

在审议海洋保护区和其他有关的划区管理工具时，提案要点应包括：

- 地理/空间说明
- 威胁/脆弱性和价值
- 与识别标准有关的生态因素
- 与区域识别标准和准则有关的科学数据
- 养护和可持续利用目标
- 相关全球、区域和部门机构的作用
- 区域内或毗邻区域的现有措施
- 区域内具体的人类活动
- 社会—经济考虑因素
- 管理计划草案
- 监测、研究和审查计划。

（二）就提案进行协商和评估

案文将规定一个就提案与相关全球、区域和部门机构，包括毗邻沿海国在内的所有国家以及其他相关利益攸关方包括科学家、业界、民间社会、传统知识拥有者和地方社区进行协调和磋商的程序。

案文还将规定对提案进行科学评估的导则。

（三）决策

案文将规定如何就划区管理工具包括海洋保护区的有关事项作出决策，包括决策人和决策依据。

案文将处理拟议划区管理工具包括海洋保护区所涉区域的毗邻沿海国的参与问题。

4.4　执行

案文将规定本文书的缔约方对于特定区域相关措施的职责。

4.5　监测和审查

案文将规定评估划区管理工具包括海洋保护区有效性以及之后采取后续行动的条款，同时注意有必要采取适应性办法。

5. 环境影响评估

5.1　进行环境影响评估的义务

根据《公约》第206条和习惯国际法，案文将规定各国有义务评估在其管辖或控制下计划开展的活动对国家管辖范围以外区域的潜在影响。

5.2　与相关文书、框架和机构的环境影响评估程序的关系

案文将规定本文书项下环境影响评估与相关法律文书和框架以及相关全球、区域和部门机构的环境影响评估程序之间的关系。

5.3　需要进行环境影响评估的活动

案文将讨论对国家管辖范围以外区域进行环境影响评估的阈值和标准。

5.4　环境影响评估程序

案文将处理环境影响评估程序的流程步骤，例如：

- 筛查
- 确定范围
- 采用现有的最佳科学资料，包括传统知识，对影响进行预测和评价
- 公告和协商

- 发布报告和向公众提供报告
- 审议报告
- 发布决策文件
- 获取资料
- 监测和审查

案文将处理环境影响评估之后的决策问题，包括一项活动是否以及在什么条件下继续开展。

案文将处理毗邻沿海国的参与问题。

5.5 环境影响评估报告的内容

案文将说明环境影响评估报告应包含的内容，例如：

- 说明计划开展的活动
- 说明可以替代计划活动的其他选择，包括非行动性选择
- 说明范围研究的结果
- 说明计划活动对海洋环境的潜在影响，包括累积影响和任何跨边界的影响
- 说明可能造成的环境影响
- 说明任何社会经济影响
- 说明避免、防止和减轻影响的措施
- 说明任何后续行动，包括监测和管理方案
- 不确定性和知识缺口
- 一份非技术摘要。

5.6 监测、报告和审查

案文将根据并遵循《公约》第 204 至 206 条规定相关义务，以确保对国家管辖范围以外区域授权开展的活动造成的影响进行监测、报告和审查。

案文将处理向毗邻沿海国提供信息的问题。

5.7　环境战略评估

案文可处理战略性环境评估问题[①]。

6. 能力建设和海洋技术转让[②]

6.1　能力建设和海洋技术转让的目标

案文将述及能力建设和海洋技术转让的目标，依照《公约》第二百六十六条第二款，通过发展和加强可能有需要和要求的国家、特别是发展中国家的能力，协助其履行该文书规定的权利和义务，从而支持实现国家管辖范围以外区域海洋生物多样性的养护和可持续利用。

案文应当承认发展中国家，特别是最不发达国家、内陆发展中国家、地理不利国和小岛屿发展中国家以及非洲沿海国家在该文书项下的特殊要求。

6.2　能力建设和海洋技术转让的类别和模式

在现有文书，例如《公约》和政府间海洋学委员会的《海洋技术转让标准和准则》的基础上，案文可以包括一份在稍后阶段制订的指示性不完全清单，列出能力建设和海洋技术转让的大类类型，例如：

• 科学和技术援助，包括有关海洋科学研究的援助，例如通过联合研究合作方案提供援助

• 教育和人力资源培训，包括采取讲习班和讨论会方式

• 数据和专门知识

案文还将提供能力建设和海洋技术转让的各种模式，包括可能采取这些模式：

---

① 可以在该文书的另一个章节，例如在划区管理工具包括海洋保护区的章节中审议这个问题。

② 能力建设和海洋技术转让可以作为专题占据专门一个章节，或者作为主流内容纳入所有其他章节。

- 由国家主导并能顺应定期评估的需求和优先事项
- 发展和加强人的能力和机构能力
- 长期且可持续
- 按照《公约》第十三和十四部分，发展各国的海洋科学和技术能力。案文将详细规定与海洋遗传资源包括惠益分享问题、划区管理工具包括海洋保护区等措施以及环境影响评估有关的合作和援助形式。

案文将作出安排，建立一个信息交换机制，以履行能力建设和海洋技术转让职能，同时考虑到其他组织的工作。[①]

6.3 供资

考虑到现有机制，案文将处理资金和资源提供问题。还可以处理有关资金和资源的持续性、可预测性和可获取性问题。

6.4 监测和审查

案文将处理对能力建设和海洋技术转让活动有效性的监测和审查问题，以及可能采取的后续行动。

四、体制安排

案文将规定体制安排，同时考虑到是否有可能利用现有的机构、制度和机制。可能的体制安排可以包括以下各项。

1. 决策机构/论坛

案文将规定一种用于决策的体制框架及其可以履行的职能。决策机构/论坛在支持文书执行方面可能履行的职能包括：

- 通过议事规则
- 审查文书的执行工作
- 有关文书执行的信息交流
- 促进为养护和可持续利用国家管辖范围以外区域海洋生物

① 信息交换机制的功能可以载于文书的单独章节，专门讨论信息交换机制（例如，见下文第五部分），或载于本节。

多样性所作的各种努力协调一致

• 促进合作与协调，包括与相关全球、区域和部门机构进行合作与协调，以养护和可持续利用国家管辖范围以外区域的海洋生物多样性

• 就文书的执行进行决策并提出建议

• 为履行职能，设立必要的附属机构

• 文书中确定的其他职能

2. 科学/技术机构

案文将规定科学咨询/信息方面的体制框架。

案文还将规定该体制框架将履行的职能，例如向文书列明的决策机构/论坛提供咨询意见以及履行决策机构/论坛确定的其他职能。

3. 秘书处

案文将规定一个履行如下秘书处职能的体制框架：

• 提供行政和后勤支持

• 应缔约国要求，报告与文书执行有关的事项以及与国家管辖范围以外区域海洋生物多样性的养护和可持续利用有关的事态发展

• 为决策机构/论坛及其可能设立的任何其他机构举办会议并提供会议服务

• 散发有关文书执行的信息

• 确保与其他有关国际机构的秘书处进行必要协调

• 按照决策机构/论坛授予的任务，协助执行本文书

• 履行文书明确规定的其他秘书处职能以及决策机构/论坛可能确定的其他职能

五、信息交换机制

案文将规定就国家管辖范围以外区域海洋生物多样性的养护和可持续利用促进相关信息交流的模式，以确保执行文书。

案文将就数据储存库或信息交换机制等各种机制作出安排。信息交换机制可能发挥的功能包括：

• 传播国家管辖范围以外区域海洋遗传资源有关研究所产生的资料、数据和知识，以及有关海洋遗传资源的其他相关资料

• 传播与划区管理工具包括海洋保护区有关的资料，例如科学数据、后续报告和主管机构作出的相关决定

• 传播关于环境影响评估的资料，例如提供一个文献中心，存储环境影响评估报告、传统知识、最佳环境管理做法和累积影响资料

• 传播能力建设和海洋技术转让相关信息，包括促进技术和科学合作的相关信息、关于研究方案、项目和举措的信息、关于能力建设和海洋技术转让有关需求和机会的信息、关于供资机会的信息

六、财政资源和财务事项

案文将处理与文书运作有关的财务事项。

七、遵守

案文将处理遵守文书方面的事项。

八、争端解决

在《联合国宪章》和《公约》的争端解决条款等现有规则基础上，案文将规定以和平方式解决争端的义务以及合作避免争端的必要性。

案文还将规定涉及文书解释或适用的争端解决模式。

九、职责和责任

案文将处理与职责和责任有关的事项。

十、审查

案文将规定定期审查文书在实现其目标方面的有效性。

十一、最后条款

案文将列明文书的最后条款。

为实现普遍参与，该文书将在这方面与《公约》的有关条款（包括涉及国际组织的条款）保持一致。

案文将解决本文书如何不妨害各国就陆地和海上争端所持立场的问题。

B 节

在人类共同财产和公海自由方面，还需要进一步讨论。

在海洋遗传资源包括分享惠益问题上，需要进一步讨论文书是否应当对海洋遗传资源的获取进行规制、这些资源的性质、应当分享何种惠益、是否处理知识产权问题、是否规定对国家管辖范围以外区域海洋遗传资源的利用进行监测。

在划区管理工具包括海洋保护区等措施方面，还需要进一步讨论最适当的决策和体制安排，以期增进合作与协调，同时避免损害现行法律文书和框架以及区域机构和（或）部门机构的授权任务。

在环境影响评估方面，还需要进一步讨论该进程由各国开展或者“国际化”的程度问题，以及文书是否应当处理战略性环境影响评估。

在能力建设和海洋技术转让方面，需要进一步讨论海洋技术转让的条款和条件。

需要进一步讨论体制安排以及国际文书建立的制度与相关全球、区域和部门机构之间的关系。还需要进一步关注的一个相关问题是如何处理监测、审查及遵守文书事项。

关于供资，需要进一步讨论所需资金的规模和是否应当设立一个财政机制。还需要进一步讨论争端解决以及职责和责任。

四、其他事项

39. 大会在第 69/292 号决议第 5 段中请秘书长设立一项特别自愿信托基金，用于协助发展中国家，特别是最不发达国家、内陆发展中国家和小岛屿发展中国家出席预备委员会会议和政府间会议，邀请会员国、国际金融机构、捐助机构、政府间组织、非政府组织以及自然人和法人向该自愿信托基金作出财政捐助。秘书处在预备委员会各届会议上通报信托基金的现况。以下各国已向自愿信托基金作出捐助：爱沙尼亚、芬兰、爱尔兰、荷兰和新西兰。

五、通过预备委员会的报告

40. 2017 年 7 月 20 日，在第 46 次会议上，主席介绍了预备委员会的报告草稿。

41. 2017 年 7 月 21 日，在第 47 次会议上，欧洲联盟及其成员国要求该报告指出，欧洲联盟及其成员国认为建议的 A 节第二 .4 部分第 3 段并不是多数代表团已形成一致意见的一个要点。

42. 在同一次会议上，预备委员会通过了经修正的报告草稿。

附件：文件一览表

A/AC. 287/2016/PC. 1/1　第一届会议议程

A/AC. 287/2016/PC. 1/L. 1　第一届会议临时议程

A/AC. 287/2016/PC. 1/L. 2　第一届会议暂定工作方案

A/AC. 287/2016/PC. 2/1　第二届会议议程

A/AC. 287/2016/PC. 2/L. 1　第二届会议临时议程

A/AC. 287/2016/PC. 2/L. 2　　第二届会议暂定工作方案
A/AC. 287/2017/PC. 3/1　　第三届会议议程
A/AC. 287/2016/PC. 3/L. 1　　第三届会议临时议程
A/AC. 287/2017/PC. 3/L. 2　　第三届会议暂定工作方案
A/AC. 287/2017/PC. 4/1　　第四届会议议程
A/AC. 287/2016/PC. 4/L. 1　　第四届会议临时议程
A/AC. 287/2017/PC. 4/L. 2　　第四届会议暂定工作方案

Report of the Preparatory Committee established by General Assembly resolution69/292: Development of an international legally binding instrument under the United Nations Convention on the Law of the Sea on the conservation and sustainable use of marine biological diversity of areas beyond national jurisdiction

Ⅰ. Introduction

1. Inits resolution 69/292 of 19 June 2015, the General Assembly decided to develop an international legally binding instrument under the United Nations Convention on the Law of the Sea (the Convention) on the conservation and sustainable use of marine biological diversity of areas beyond national jurisdiction. To that end, the Assembly decided to establish, prior to holding an intergovernmental conference, a preparatory committee, open to all States Members of the United Nations, members of the specialized agencies and parties to t he Convention, with others invited as observers in accordance with past practice of the United Nations, to make substantive recommendations to the Assembly on the elements of a draft text of an international legally binding instrument under the Convention, taking into account the various reports of the Co-Chairs on the work of the Ad Hoc Open-ended Informal Working Group

to study issues relating to the conservation and sustainable use of marine biological diversity beyond areas of national jurisdiction.①

2. The General Assembly also decided that the Preparatory Committee would start its work in 2016 and, by the end of 2017, report to it on its progress, and that the Assembly, before the end of its seventy-second session, and taking into account the aforementioned report of the Preparatory Committee, would decide on the convening and on the starting date of an intergovernmental conference, under the auspices of the United Nations, to consider the recommendations of the Preparatory Committeeon the elements and to elaborate the text of an international legally binding instrument under the Convention.

3. The General Assembly recognized the desirability that any legally binding instrument relating to marine biological diversity of areas beyond national jurisdiction under the Convention would secure the widest possible acceptance, and for that reason, decided that the Preparatory Committee should exhaust every effort to reach agreement on substantive matters by consensus. It also recognized the importance of proceeding efficiently in the Preparatory Committee on the development of the elements of a draft text of an international legally binding instrument under the Convention, and recognized further that any elements where consensus was not attained, even after exhausting every effort, might also be included in a section of the recommendations of the Preparatory Committee to the General Assembly.

4. The General Assembly decided that negotiations should address the topics identified in the package agreed in 2011 (see resolution 66/

① See A/61/65, A/63/79 and Corr. 1, A/65/68, A/66/119, A/67/95, A/68/399, A/69/82, A/69/177 and A/69/780.

231）, namely the conservation and sustainable use of marine biological diversity of areas beyond national jurisdiction, in particular, together and as a whole, marine genetic resources, including questions on the sharing of benefits, measures such as area- based management tools, including marine protected areas, environmental impact assessments and capacity-building and the transfer of marine technology.

5. It further recognized that the process should not undermine existing relevant legal instruments and frameworks and relevant global, regional and sectoral bodies, and that neither participation in the negotiations nor their outcome might affect the legal status of non-parties to the Convention or any other related agreements with regard to those instruments, or the legal status of parties to the Convention or any other related agreements with regard to those instruments.

6. Inaccordance with paragraph 6 of resolution 69/292, substantive secretariat support was provided to the Preparatory Committee by the Division for Ocean Affairs and the Law of the Sea of the Office of Legal Affairs of the Secretariat.

Ⅱ. Organizational matters

A. Sessionsof the Preparatory Committee

7. Pursuantto General Assembly resolution 69/292, in which the Assembly decided that the Preparatory Committee should meet for no less than two sessions of a duration of 10 working days each in 2016 and 2017, the Secretary-General convened the first and second sessions of the Preparatory Committee from 28 March to 8 April 2016 and from 26 August to 9 September 2016, respectively, at United Nations Head-

quarters. Pursuant to resolution 71/257, the third session of the Preparatory Committee was convened by the Secretary- General from 27 March to 7 April 2017 and the fourth session from 10 to 21 July 2017 at United Nations Headquarters.

B. Election of officers

8. Bya letter dated 4 September 2015 addressed to Member States, Sam Kahamba Kutesa, President of the General Assembly at its sixty-ninth session, in accordance with paragraph 1 (d) of resolution 69/292, appointed Eden Charles, Deputy Permanent Representative of Trinidad and Tobago and Chargé d' affaires a. i. of the Permanent Mission of Trinidad and Tobago to the United Nations, as Chair of the Preparatory Committee.

9. Atits first session, in accordance with paragraph 1 (e) of resolution 69/292, in which the General Assembly decided that the Preparatory Committee should elect a bureau consisting of two members from each regional group, and that those members should assist the Chair on procedural matters in the general conduct of his or her work, the Preparatory Committee elected a bureau consisting of the following members: Mohammed Atlassi (Morocco), Thembile Elphus Joyini (South Africa), Ma Xinmin (China), Kaitaro Nonomura (Japan), Konrad Marciniak (Poland), Maxim V. Musikhin (Russian Federation), Javier Gorostegui Obanoz (Chile), Gina Guillén-Grillo (Costa Rica), Antoine Misonne (Belgium) and Giles Norman (Canada).

10. Atits second session, the Preparatory Committee elected Jun Hasebe (Japan) and Catherine Boucher (Canada) as members of the bureau, replacing Kaitaro Nonomura and Giles Norman, who had re-

signed from their positions as bureau members. The Preparatory Committee further elected Margo Deiye (Nauru) to serve as a member of the bureau from 28 October 2016 onwards in light of an agreement reached in the Asia-Pacific Group to share membership on the bureau.

11. Bya letter dated 24 January 2017, Peter Thomson, President of the General Assembly at its seventy-first session, informed Member States that Eden Charles had indicated that he would no longer be in a position to serve as Chair of the Preparatory Committee. He further stated that, following consultations with Member States, he had, in accordance with paragraph 1 (d) of resolution 69/292, appointed Carlos Sergio Sobral Duarte, Deputy Permanent Representative of Brazil to the United Nations, as Chair of the Preparatory Committee.

12. Atits third session, in accordance with paragraph 1 (e) of General Assembly resolution 69/292, and in light of an agreement reached in the Group of Latin American and Caribbean States, the Preparatory Committee elected Pablo Adrían Arrocha Olabuenaga (Mexico) and José Luis Fernandez Valoni (Argentina) as members of the bureau in replacement of Javier Gorostegui Obanoz and Gina Guillén-Grillo. In the light of information received from Malaysia as Chair of the Asia-Pacific Group and in accordance with the agreement reached in that group, the Preparatory Committee further elected Jun Hasebe (Japan) to serve as a member of the bureau from 28 May 2017, replacing Ma Xinmin, whose resignation from the bureau was to take effect on 27 May 2017.

C. Documentation

13. Inits resolution 69/292, the General Assembly recognized that, with respect to documentation, any documents of the Preparatory Committee other than the agenda, the programme of work and the report of

the Preparatory Committee should be considered informal working documents. The list of official documents before the Preparatory Committee at its sessions is annexed to the present report.

14. Inaddition, with a view to assisting the proceedings, the Chair prepared a number of informal documents under his responsibility (see paras. 21, 26 and 32 below), including the Chair's overview of the first, second and third sessions and a streamlined non－paper on elements of a draft text of an international legally binding instrument under the United Nations Convention on the Law of the Sea on the conservation and sustainable use of marine biological diversity of areas beyond national jurisdiction. ①

15. Atthe invitation of the Chair, delegations also submitted views on the elements of a draft text, which were made available on the website of the Division for Ocean Affairs and the Law of the Sea. ②

D. Proceedings ofthe sessions of the Preparatory Committee

16. Inits resolution 69/292, the General Assembly decided that, except as provided for in subparagraph 1 (i) of that resolution, in which the Assembly recognized the importance of proceeding efficiently in the Preparatory Committee on the development of the elements of a draft text of an international legally binding instrument under the Convention, and recognized further that any elements where consensus was not attained, even after exhausting every effort, might also be included in a section of the recommendations of the Preparatory Committee, the

① Available from www. un. org/depts/los/biodiversity/prepcom. htm.

② Ibid.

rules relating to the procedure and the established practice of the committees of the General Assembly would apply to the procedure of the Preparatory Committee, and that, for the meetings of the Preparatory Committee, the participation rights of the international organization that was a party to the Convention would be as in the Meeting of States Parties to the Convention and that that provision would constitute no precedent for all meetings to which General Assembly resolution 65/276 of 3 May 2011 was applicable.

1. First session

17. Atthe 1st meeting of the Preparatory Committee, on 28 March 2016, the Under－Secretary－General for Legal Affairs and United Nations Legal Counsel made a statement. The Preparatory Committee adopted the agenda of the session as contained in document AC. 287/2016/PC. 1/L. 1 and agreed to proceed on the basis of the provisional programme of work contained in document A/AC. 287/2016/PC. 1/L. 2.

18. ThePreparatory Committee held 15 plenary meetings at its first session. Representatives of 99 States Members of the United Nations, two non－member States, five United Nations programmes, funds and offices, four specialized agencies and related organizations of the United Nations system, eight intergovernmental organizations and 17 non－governmental organizations attended the meeting.

19. Atits plenary meetings the Preparatory Committee heard general statements and considered the following issues: scope of an international legally binding instrument and its relationship with other instruments; guiding approaches and principles of an international legally binding instrument; marine genetic resources, including questions on the sharing of benefits; measures such as area－based management tools, including

marine protected areas; environmental impact assessments; and capacity-building and the transfer of marine technology. The plenary also discussed and approved a road map for the second session.

20. Informal workinggroup sessions were also convened and facilitated as follows: Carlos Duarte (Brazil) for the informal working group on marine genetic resources, including questions on the sharing of benefits; John Adank (New Zealand) for the informal working group on measures such as area-based management tools, including marine protected areas; René Lefeber (Netherlands) for the informal working group on environmental impact assessments; and Rena Lee (Singapore) for the informal working group on capacity-building and the transfer of marine technology.

21. Inaccordance with the road map discussed and approved by the plenary, following the first session, the Chair prepared an overview of the session. The Chair also prepared indicative suggestions of clusters of issues and questions to assist further discussions in the informal working groups at the second session of the Preparatory Committee. ①

2. Second session

22. Atthe 16th meeting of the Preparatory Committee, on 26 August 2016, the Assistant Secretary-General for Legal Affairs made a statement. The Preparatory Committee adopted the agenda as contained in document A/AC. 287/2016/PC. 2/L. 1 andagreed to proceed on the basis of the provisional programme of work contained in document A/AC. 287/2016/PC. 2/L. 2.

23. ThePreparatory Committee held 13 plenary meetings at its second session. Representatives of 116 States Members of the United Na-

① Available from www. un. org/depts/los/biodiversity/prepcom. htm.

tions, three non－member States, six United Nations programmes, funds and offices, and five specialized agencies and related organizations of the United Nations system of the United Nations, nine intergovernmental organizations and 22 non－governmental organizations attended the session.

24. Atits plenary meetings the Preparatory Committee considered the following issues: marine genetic resources, including questions on the sharing of benefits; measures such as area－based management tools, including marine protected areas; environmental impact assessments; capacity－building and the transfer of marine technology; and cross－cutting issues. The plenary also discussed and approved a road map for the third session.

25. Informal workinggroup sessions were also convened and facilitated as follows: Eden Charles (Trinidad and Tobago) ① for the informal working group on marine genetic resources, including questions on the sharing of benefits; John Adank (New Zealand) for the informal working group on measures such as area－ based management tools, including marine protected areas; René Lefeber (Netherlands) for the informal working group on environmental impact assessments ; Rena Lee (Singapore) for the informal working group on capacity－building and the transfer of marine technology; and the Chair of Preparatory Committee, Eden Charles (Trinidad and Tobago), for the informal working group on cross－cutting issues.

26. Inaccordance with the road map discussed and approved by the plenary, following the second session, the Chair prepared an overview

① The Chair facilitated the informal working group in the light of the unavailability of Carlos Duarte.

of the session. The Chair also prepared the Chair's non - paper on elements of a draft text of an international legally-binding instrument under the United Nations Convention on the Law of the Sea and a supplement to that paper. [1]

3. Third session

27. Atthe 29th meeting of the Preparatory Committee, on 27 March 2017, the Under - Secretary - General for Legal Affairs and United Nations Legal Counsel made a statement. The Preparatory Committee adopted the agenda as contained in document A/AC. 287/2017/PC. 3/L. 1 and agreed to proceed on the basis of the provisional programme of work contained in document A/AC. 287/2017/PC. 3/L. 2.

28. ThePreparatory Committee held nine plenary meetings at its third session. Representatives of 147 States Members of the United Nations, two non-member States, five United Nations programmes, funds and offices, four specialized agencies and related organizations of the United Nations system, 14 intergovernmental organizations and 19 non-governmental organizations attended the session.

29. TheCo-Chair of the Ad Hoc Working Group of the Whole on the Regular Process for Global Reporting and Assessment of the State of the Marine Environment, including Socioeconomic Aspects, introduced the advance and unedited text of the Technical Abstract of the First Global Integrated Marine Assessment on the Conservation and Sustainable Use of Marine Biological Diversity of Areas Beyond National Jurisdiction, under item 7, "Other matters". Under this item, Peter Thomson, President of the General Assembly at its seventy - first session, also addressed the Preparatory Committee.

---

[1] Available from www. un. org/depts/los/biodiversity/prepcom. htm.

30. Atits plenary meetings the Preparatory Committee considered the following issues: marine genetic resources, including questions on the sharing of benefits; measuressuch as area-based management tools, including marine protected areas; environmental impact assessments; capacity-building and the transfer of marine technology; and cross-cutting issues. The plenary also discussed and approved a road map for the fourth session.

31. Informal working group sessionswere also convened and facilitated as follows: Janine Elizabeth Coye-Felson (Belize)① for the informal working group on marine genetic resources, including questions on the sharing of benefits; Alice Revell (New Zealand)② for the informal working group on measures such as area-based management tools, including marine protected areas; René Lefeber (the Netherlands) for the informal working group on environmental impact assessments; Rena Lee (Singapore) for the informal working group on capacity-building and the transfer of marine technology; and the Chair of the Preparatory Committee, Carlos Duarte, for the informal working group on cross-cutting issues.

32. Inaccordance with the road map discussed and approved by the plenary, following the third session, the Chair prepared an overview of the session. The Chair also prepared indicative suggestions to assist the Preparatory Committee in developing recommendations to the General Assembly on the elements of a draft text of an international legally

---

① In replacement of Carlos Duarte (Brazil), owing to his new role as Chair of the Preparatory Committee.

② In replacement of John Adank (New Zealand), who had informed the Chair that he was no longer in a position to serve as facilitator.

binding instrument under the United Nations Convention on the Law of the Sea, as well as a streamlined non-paper on elements of a draft text of an international legally-binding instrument under the United Nations Convention on the Law of the Sea.[①]

4. Fourth session

33. Atthe 38th meeting of the Preparatory Committee, on 10 July 2017, the Under-Secretary-General for Legal Affairs and United Nations Legal Counsel made a statement. The Committee adopted the provisional agenda as contained in document A/AC. 287/2017/PC. 4/L. 1 and agreed to proceed on the basis of the proposed programme of work contained in document A/AC. 287/2017/PC. 4/L. 2.

34. ThePreparatory Committee held 10 plenary meetings at its fourth session. Representatives of 131 States Members of the United Nations, two non-member States, two United Nations programmes, funds and offices, nine specialized agencies and related organizations of the United Nations system, 10 intergovernmental organizations and 23 non-governmental organizations attended the session.

35. Atits plenary meetings the Preparatory Committee heard general statements and considered the development of substantive recommendations on the elements of a draft text of an international legally binding instrument under the United Nations Convention on the Law of the Sea (see para. 38 below). It also considered the report of the Preparatory Committee (see para. 40 below).

36. Duringthe first week, informal working group sessions were also convened and facilitated as follows: Janine Elizabeth Coye-Felson (Belize) for the informal working group on marine genetic resources, inclu-

① Available from www. un. org/depts/los/biodiversity/prepcom. htm.

ding questions on the sharing of benefits; Alice Revell (New Zealand) for the informal working group on measures such as area-based management tools, including marine protected areas; René Lefeber (Netherlands) for the informal working group on environmental impact assessments; Rena Lee (Singapore) for the informal working group on capacity- building and the transfer of marine technology; and the Chair of the Preparatory Committee, Carlos Duarte (Brazil), for the informal working group on cross-cutting issues.

37. Duringthe plenary sessions in the second week, many delegations proposed to include as part of the substantive recommendations to the General Assembly that an intergovernmental conference be convened in 2018. Some delegations also proposed that the conference should have at least four rounds of negotiations, each held for a period of two weeks during 2018 and 2019, with full conference services. Some delegations suggested that the rules of procedure of the General Assembly should apply mutatis mutandis to the conference. Other delegations stressed that discussions on whether, when and how to convene an intergovernmental conference should be left to the General Assembly, and that the substantive recommendations of the Preparatory Committee should not contain any suggestions in that regard in order not to prejudge such discussions in the Assembly. A delegation was of the view that additional sessions of the Preparatory Committee might be needed before moving towards an intergovernmental conference.

Ⅲ. Recommendations of the Preparatory Committee

38. Atits 47th meeting, on 21 July 2017, the Preparatory

Committee adopted, by consensus, the following recommendations.

The Preparatory Committee, having met in accordance with General Assembly resolution69/292 of 19 June 2015, recommends to the General Assembly:

(a) That the elements contained in sections A and B below be considered with a view to the development of a draft text of an international legally binding instrument under the United Nations Convention on the Law of the Sea on the conservation and sustainable use of marine biological diversity of areas beyond national jurisdiction. Sections A and B do not reflect consensus. Section A includes non-exclusive elements that generated convergence among most delegations. Section B highlights some of the main issues on which there is divergence of views. Sections A and B are for reference purposes because they do not reflect all options discussed. Both sections are without prejudice to the positions of States during the negotiations;

(b) That the General Assembly take a decision, as soon as possible, on the convening of an intergovernmental conference, under the auspices of the United Nations, to consider the recommendations of the Preparatory Committee on the elements and to elaborate the text of an international legally binding instrument under the Convention.

Section A

I. Preambular elements

The text would set out broad contextual issues, such as:

• Adescription of the considerations that led to the development of the instrument, including key concerns and issues

• Recognitionof the central role of the Convention and the role of other existing relevant legal instruments and frameworks and relevant global, regional and sectoral bodies for the conservation and sustainable use of marine biological diversity of areas beyond national jurisdiction

• Recognitionof the need to enhance cooperation and coordination for the conservation and sustainable use of marine biological diversity of areas beyond national jurisdiction

• Recognitionof the need for assistance so that developing countries, in particular geographically disadvantaged States, least developed countries, landlocked developing countries and small island developing States, as well as coastal African States, can participate effectively in the conservation and sustainable use of marine biological diversity of areas beyond national jurisdiction

• Recognitionof the need for the comprehensive global regime to better address the conservation and sustainable use of marine biological diversity of areas beyond national jurisdiction

• Anexpression of conviction that an agreement for the implementation of the relevant provisions of the Convention would best serve these purposes and contribute to the maintenance of international peace and security

• Anaffirmation that matters not regulated by the Convention, its Implementing Agreements or the instrument continue to be governed by the rules and principles of general international law.

Ⅱ. General elements

1. Use of terms①

The text would provide definitions of key terms, bearing in mind the need for consistency with those contained in the Convention and

① Some specific definitions of relevance to only one part of the instrument could be included under the respective parts.

other relevant legal instruments and frameworks.

2. Scope of application

2.1 Geographical scope

The text would state thatthe instrument applies to areas beyond national jurisdiction.

Itwould state that the rights and jurisdiction of coastal States over all areas under their national jurisdiction, including the continental shelf within and beyond 200 nautical miles and the exclusive economic zone, shall be respected.

2.2 Mqterial scope

The text would address the conservation and sustainable use of marine biological diversity of areas beyond national jurisdiction, in particular, together and as a whole, marine genetic resources, including questions on the sharing of benefits, measures such as area-based management tools, including marine protected areas, environmental impact assessments and capacity-building and the transfer of marine technology.

It could set out exclusions from the scope of application of the instrument, and address, consistent with the Convention, issues relating to sovereign immunity.

3. Objective (s)

The text wouldset out that the objective of the instrument is to ensure the conservation and sustainable use of marine biological diversity of areas beyond national jurisdiction through effective implementation of the Convention.

Itcould further set out additional objectives, if agreed, such as furthering international cooperation and coordination, to ensure the achievement of the overall objective of conservation and sustainable use of marine biological diversity of areas beyond national jurisdiction.

4. Relationship to the Convention and otherinstruments and frameworks and relevant global, regional and sectoral bodies

Withregard to the relationship to the Convention, the text would state that nothing in the instrument shall prejudice the rights, jurisdiction and duties of States under.

the Convention. It would further state that the instrument shall be interpreted and applied in the context of and in a manner consistent with the Convention.

The text would state that the instrument would promote greater coherence with and complement existing relevant legal instruments and frameworks and relevant global, regional and sectoral bodies. It would also state that the instrument should be interpreted and applied in a manner which would not undermine these instruments, frameworks and bodies.

The text couldrecognize that the legal status of non-parties to the Convention or any other related agreements with regard to those instruments would not be affected.

Ⅲ. Conservation and sustainable use of marine biological diversity of areasbeyond national jurisdiction

1. General principles and approaches①

The text would set outthe general principles and approaches guiding the conservation and sustainable use of marine biological diversity of areas beyond national jurisdiction.

Possible general principles and approaches could include:

- Respectfor the balance of rights, obligations and interests en-

① Some of these principles and approaches would be included in a separate article and some in the preamble.

shrined in the Convention

• Dueregard as reflected in relevant provisions of the Convention

• Respectfor the rights and jurisdiction of coastal States over all areas under their national jurisdiction, including the continental shelf within and beyond 200 nautical miles and the exclusive economic zone

• Respectfor the sovereignty and territorial integrity of all States

• Use of marinebiological diversity of areas beyond national jurisdiction for peaceful purposes only

• Promotionof both the conservation and sustainable use of marine biological diversity of areas beyond national jurisdiction

• Sustainable development

• International cooperation and coordination, at all levels, including north-south, south-south, and triangular cooperation

• Relevant stakeholders engagement

• Ecosystem approach

• Precautionary approach

• Integrated approach

• Science-based approach, using the best available scientific information and knowledge, including traditional knowledge

• Adaptive management

• Building resilienceto the effects of climate change

• Dutynot to transform one type of pollution into another consistent with the Convention

• Polluter-pays principle

• Public participation

• Transparencyand availability of information

• Special requirementsof small islands developing States and least developed countries, including avoiding transferring, directly or indirectly, a disproportionate burden of conservation action onto developing countries

• Good faith.

2. International cooperation

The text would set out the obligation of States to cooperate for the conservation and sustainable use of marine biological diversity of areas beyond national juri sdiction, and elaborate on the content and modalities of this obligation.

3. Marine genetic resources, including questions on the sharing of benefits

3.1 Scope①

The text would set out the geographical and material scope of application of this section of the instrument.

3.2 Accessand benefit-sharing

3.2.1 Access

The text would address access.

3.2.2 Sharingof benefits

(i) *Objectives*

The text would set out that the objectives of benefit-sharing are:

• Contributingto the conservation and sustainable use of marine biological diversity of areas beyond national jurisdiction

• Building capacityof developing countries to access and use marine genetic resources of areas beyond national jurisdiction.

It could further set out additional objectives, if agreed.

(ii) *Principlesand approaches guiding benefit-sharing*②

---

① The scope could also be included in an overarching section on scope at the beginning of the instrument (see, for example, part II. 2 above).

② The principles could also be included in an overarching section on principles and approaches at the beginning of the instrument (see, for example, part III. 1 above).

The text wouldset out the principles and approaches guiding benefit-sharing, such as:

- Beingbeneficial to current and future generations
- Promotingmarine scientific research and research and development.

(iii) *Benefits*

The text would set out the types of benefits that could be shared.

(iv) *Benefit-sharing modalities*

The text would set out modalities for the sharing of benefits, taking into account existing instruments and frameworks, for example it could make provision for a clearing-house mechanism with regard to the sharing of benefits. ①

3.2.3 Intellectual property rights

The text couldset out the relationship between the instrument and intellectual property rights.

3.3 Monitoringof the utilization of marine genetic resources of areas beyond national jurisdiction

The text couldaddress monitoring of the utilization of marine genetic resources of areas beyond national jurisdiction.

4. Measures such as area-based management tools, including marine protected areas

4.1 Objectivesof area-based management tools, including marine protected areas

The text would set out objectives of area-based management tools, including marine protected areas, in areas beyond national jurisdiction

① The functions of a clearing-house mechanism could be set out in a stand-alone section of the instrument dedicated to the clearing-house mechanism (see, for example, part V below) or in this section.

for the conservation and sustainable use of marine biological diversity.

4.2 Relationshipto measures under relevant instruments, frameworks and bodies

The text wouldset out the relationship between measures under the instrument and measures under existing relevant legal instruments and frameworks and relevant global, regional and sectoral bodies, for the purpose of coherence and coordination of efforts.

The text would affirmthe importance of enhanced cooperation and coordination between relevant legal instruments and frameworks and relevant global, regional and sectoral bodies, with regard to area-based management tools, including marine protected areas, without prejudice to their respective mandates.

The text would also address the relationship between measures under the instrument and those established by adjacent coastal States, including issues of compatibility, without prejudice to the rights of coastal States.

4.3 Processin relation to area-based management tools, including marine protected areas

Taking into account the various types of area-based management tools, including marine protected areas, the text would set out the process in relation to area-based management tools, including marine protected areas, and the relevant roles and responsibilities, on the basis of the approach that will be developed.

4.3.1 Identificationof areas

The text wouldset out that the process for identification of areas within whic h protection may be required would be based on the best available scientific information, standards and criteria, including:

- Uniqueness

- Rarity
- Special importancefor life history stages of species
- Importancefor threatened, endangered or declining species and/or habitats
- Vulnerability
- Fragility
- Sensitivity
- Biological productivity
- Biological diversity
- Representativeness
- Dependency
- Naturalness
- Connectivity
- Ecological processes
- Economicand social factors.

4.3.2 Designation process

(i) *Proposal*

The text would contain provisions on proposals related to area-based management tools, including marine protected areas.

When considering marine protectedareas, and other area-based management tools where relevant, the elements of the proposal should include:

- Geographic/spatial description
- Threats/vulnerabilitiesand values
- Ecological factors relatedto identification criteria
- Scientificdata concerning the standards and criteria for the identification of the area
- Conservationand sustainable use objectives

- The role ofrelevant global, regional and sectoral bodies
- Existingmeasures in the area or areas adjacent to it
- Specific human activitiesin the area
- Socio-economic considerations
- Adraft management plan
- Monitoring, researchand review plan.

(ii) *Consultationon and assessment of the proposal*

The text would set out a process for coordination and consultations on the proposal with relevant global, regional and sectoral bodies, all States, including adjacent coastal States, and other relevant stakeholde-rs, including scientists, industry, civil society, traditional knowledge holders and local communities.

It would also set out guidance for a scientific assessment of the proposal.

(iii) *Decision-making*

The text would set out how decisions on matters related to area-based management tools, including marine protected areas, would be made, including who would make the decision and on what basis.

The text wouldaddress the question of the involvement of coastal States adjacent to an area for which area-based management tools, including marine protected areas, are proposed.

4.4 Implementation

The text would set out the responsibility of Parties to the instrument related to the measures for a particular area.

4.5 Monitoringand review

The text would set outprovisions for assessing the effectiveness of area-based management tools, including marine protected areas, and subsequent follow-up action, bearing in mind the need for an adaptive approach.

5. Environmental impact assessments

5.1 Obligationto conduct environmental impact assessments

Drawing from article206 of the Convention and customary international law, the text would set out the obligation for States to assess the potential effects of planned activities under their jurisdiction or control in areas beyond national jurisdiction.

5.2 Relationshipto environmental impact assessment processes under relevant instruments, frameworks and bodies

The text would set outthe relationship to environmental impact assessment processes under relevant legal instruments and frameworks and relevant global, regional and sectoral bodies.

5.3 Activitiesfor which an environmental impact assessment is required

The text would address the thresholds and criteria for undertaking environmental impact assessments in respect of areas beyond national jurisdiction.

5.4 Environmental impact assessment process

The text would address the procedural steps of an environmental impact assessment process, such as:

- Screening
- Scoping
- Impact predictionand evaluation, using the best available scientific information, including traditional knowledge
- Publicnotification and consultation
- Publicationof reports and public availability of reports
- Considerationof reports
- Publicationof decision-making documents
- Accessto information

• Monitoringand review.

The text would address decision-making following the environmental impact assessment, including on whether an activity would proceed or not and under which conditions.

The text would address the question of involvement of adjacent coastal States.

5.5 Contentof environmental impact assessment reports

The text would address the required content of environmental impact assessment reports, such as:

• Descriptionof the planned activities

• Descriptionof reasonable alternatives to the planned activities, including non-action alternatives

• Descriptionof scoping results

• Descriptionof the potential effects of the planned activities on the marine environment, including cumulative impacts and any transboundary impacts

• Descriptionof the environment likely to be affected

• Descriptionof any socio-economic impacts

• Descriptionof any measures for avoiding, preventing and mitigating impacts

• Descriptionof any follow-up actions, including any monitoring and management programmes

• Uncertaintiesand gaps in knowledge

• Anon-technical summary.

5.6 Monitoring, reportingand review

Basedon and consistent with articles 204 to 206 of the Convention, the text would set out the obligation to ensure that the impacts of authorized activities in areas beyond national jurisdiction are monitored, reported and reviewed.

The text would address the question of information to adjacent coastal States.

5.7　Strategic environmental assessments

The text could address strategic environmental assessments.①

6. Capacity-building and transfer of marine technology②

6.1　Objectivesof capacity-building and transfer of marine technology

The text wouldaddress the objectives of capacity-building and the transfer of marine technology, in supporting the achievement of conservation and sustainable use of marine biological diversity of areas beyond national jurisdiction by developing and strengthening the capacity of States which may need and request it, particularly developing States, in accordance with article 266, paragraph 2 of the Convention, to assist them to fulfil their rights and obligations under the instrument.

The textshould recognize the special requirements under the instrument of developing countries, in particular the least developed countries, landlocked developing countries, geographically disadvantaged States and small island developing States, as well as coastal African States.

6.2　Types of and modalities for capacity-building and transfer of marine technology

Drawingon existing instruments, such as the Convention and the Criteria and Guidelines on Transfer of Marine Technology of the Intergovernmental Oceanographic Commission, the text could include an indica-

① This could be considered in a different section of the instrument, for example, in the section on area-based management tools, including marine protected areas.

② Capacity-building and transfer of marine technology could feature in a dedicated section or be mainstreamed across the other sections.

tive, non-exhaustive list, which could be developed at a later stage, of broad categories of types of capacity-building and transfer of marine technology, such as:

- Scientificand technical assistance, including with regard to marine scientific research for example through joint research cooperation programmes
- Educationand training of human resources, including through workshops and seminars
- Dataand specialized knowledge.

The text would also provide modalities for capacity-building and transfer of marine technology, including possibly for such modalities to:

- Becountry-driven and responsive to periodically assessed needs and priorities
- Developand strengthen human and institutional capacities
- Be longterm and sustainable and
- Developmarine scientific and technological capacity of States in accordance with Parts XIII and XIV of the Convention.

The text wouldelaborate on forms of cooperation and assistance in relation to marine genetic resources, including questions on the sharing of benefits, measures such as area-based management tools, including marine protected areas, and environmental impact assessments.

Itwould make provision for a clearing-house mechanism to perform functions with regard to capacity-building and transfer of marine technology, taking into account the work of other organizations. ①

---

① The functions of a clearing-house mechanism could be set out in a stand-alone section of the instrument dedicated to the clearing-house mechanism (see, for example, part V below) or in this section.

6.3　Funding

Takinginto account existing mechanisms, the text would address provision of funding and resources. Related questions on the sustainability, predictability and accessibility of such funding and resources could also be addressed.

6.4　Monitoringand review

The text wouldaddress the issue of monitoring and review of the effectiveness of capacity-building and transfer of marine technology activities, and possible follow-up action.

Ⅳ. Institutional arrangements

The text would set outinstitutional arrangements, taking into account the possibility of using existing bodies, institutions and mechanisms. Possible institutional arrangements could include the following.

1. Decision-making body/forum

The text wouldset out an institutional framework for decision-making, as well as the functions that could be performed.

Possible functionsthat a decision - making body/forum would perform in support of the implementation of the instrument could include:

- Adoptingits rules of procedure
- Reviewing implementationof the instrument
- Exchangeof information relevant to the implementation of the instrument
- Promoting coherenceamong efforts towards the conservation and sustainable use of marine biological diversity of areas beyond national jurisdiction
- Promoting cooperationand coordination, including with relevant global, regional and sectoral bodies towards the conservation and sus-

tainable use of marine biological diversity of areas beyond national jurisdiction

- Making decisionsand recommendations related to the implementation of the instrument
- Establishing subsidiary bodiesas necessary for the performance of its functions
- Other functions identifiedin the instrument.

2. Scientific/technical body

The text would set out an institutional framework for scientific advice/information.

It would also set out the functions that such an institutional framework would perform, such as providing advice to the decision-making body/forum specified in the instrument and such other functions as may be determined by the decision- making body/forum.

3. Secretariat

The text wouldset out an institutional framework for secretariat functions, such as:

- Providing administrativeand logistical support
- Reportingto States parties on matters related to the implementation of the instrument and developments related to the conservation and sustainable use of marine biological diversity of areas beyond national jurisdiction as requested by the Parties
- Conveningand servicing the meetings of the decision-making body/forum, and any other bodies as may be established by the decision-making body/forum
- Circulating information relatingto the implementation of the instrument

• Ensuringthe necessary coordination with the secretariats of other relevant international bodies

• Providing assistancefor the implementation of the instrument as mandated by the decision-making body/forum

• Performingother secretariat functions specified in the instrument and such other functions as may be determined by the decision-making body/forum.

V. Clearing-house mechanism

The text wouldset out modalities to facilitate the exchange of information relevant to the conservation and sustainable use of marine biological diversity of areas beyond national jurisdiction for the implementation of the instrument.

Itwould make provision for mechanisms such as data repositories or a clearing - house mechanism.

Possible functionsof a clearing-house mechanism could include:

• Disseminationof information, data and knowledge resulting from research relating to marine genetic resources of areas beyond national jurisdiction, information on traditional knowledge associated with marine genetic resources of areas beyond national jurisdiction, as well as other relevant information related to marine genetic resources

• Disseminationof information relating to area-based management tools, including marine protected areas, such as scientific data, follow-up reports and related decisions taken by competent bodies

• Disseminationof information on environmental impact assessments, such as by providing a central repository for reports of environmental impact assessments, traditional knowledge, best environmental management practices and cumulative impacts

• Disseminationof information relating to capacity-building and transfer of marine technology, including facilitation of technical and scientific cooperation; information on research programmes, projects and initiatives; information on needs related to capacity - building and transfer of marine technology and available opportunities; and information on funding opportunities.

Ⅵ. Financial resources and issues

The text would address financial issues relating to the operation of the instrument.

Ⅶ. Compliance

The text would address issues of compliance.

Ⅷ. Settlement of disputes

Drawingon existing dispute settlement provisions, such as those of the Charter of the United Nations and the Convention, the text would set out the obligation to settle disputes by peaceful means as well as the need to cooperate to prevent disputes.

It would also set out the modalities for settling disputes concerning the interpretation or application of the instrument.

Ⅸ. Responsibility and liability

The text could address issues relating to responsibility and liability.

Ⅹ. Review

The text could set out that the effectiveness of the instrument in achieving its objectives would be periodically reviewed.

Ⅺ. Final clauses

The text would set out the final clauses of the instrument.

Witha view to achieving universal participation, the instrument would be consistent with the relevant provisions of the Convention on this matter, including regarding international organizations.

The textcould address the issue of how not to prejudice the positions of States on land and maritime disputes.

Section B

With regard to the common heritage of mankind and the freedom of the high seas, further discussions are required.

Withregard to marine genetic resources, including the question of the sharing of benefits, further discussions are required on whether the instrument should regulate access to marine genetic resources; the nature of these resources; what benefits should be shared; whether to address intellectual property rights; and whether to provide for the monitoring of the utilization of marine genetic resources of areas beyond national jurisdiction.

Withregard to measures such as area-based management tools, including marine protected areas, further discussions are required on the most appropriate decision-making and institutional set up, with a view to enhancing cooperation and coordination, while avoiding undermining existing legal instruments and frameworks and the mandates of regional and/or sectoral bodies.

Withregard to environmental impact assessments, further discussions are required on the degree to which the process should be conducted by States or be "internationalized", as well as on whether the instrument should address strategic environmental impact assessments.

With regard to capacity-building and the transfer of marine technology, further discussions are required on the terms and conditions for the transfer of marine technology.

Furtherdiscussions are required on institutional arrangements and the relationship between the institutions established under an international instrument and relevant global, regional and sectoral bodies. A related issue that would also require further attention is how to address monitoring, review and compliance.

Withrespect to funding, further discussions are required on the scope of the financial resources required and whether a financial mechanism should be established.

Furtherdiscussions are also required on settlement of disputes and responsibility and liability.

Ⅳ. Other matters

39. Inparagraph 5 of its resolution 69/292, the General Assembly requested the Secretary – General to establish a special voluntary trust fund for the purpose of assisting developing countries, in particular the least developed countries, landlocked developing countries and small island developing States, in attending the meetings of the Preparatory Committee and the intergovernmental conference and invited Member States, international financial institutions, donor agencies, intergovernmental organizations, non – governmental organizations and natural and juridical persons to make financial contributions to the voluntary trust fund. The Secretariat informed the Preparatory Committee of the status of the trust fund at sessions of the Preparatory Committee. The following States made contributions to the voluntary trust fund: Estonia, Finland, Ireland, the Netherlands and New Zealand.

Ⅴ. Adoption of thereport of the Preparatory Committee

40. Atthe 46th meeting, on 20 July 2017, the Chair introduced the

draft report of the Preparatory Committee.

41. Atthe 47th meeting, on 21 July 2017, the European Union and its member States requested that the report indicate that, in their view, the third paragraph of part II. 4 of section A of the recommendations is not an element that generated convergence among most delegations.

42. Atthe same meeting, the Preparatory Committee adopted its draft report as amended.

Annex: List of documents

A/AC. 287/2016/PC. 1/1　Agenda of the first session

A/AC. 287/2016/PC. 1/L. 1　Provisional agenda of the first session

A/AC. 287/2016/PC. 1/L. 2　Provisional programme of work for the first session

A/AC. 287/2016/PC. 2/1　Agenda of the second session

A/AC. 287/2016/PC. 2/L. 1　Provisional agenda of the second session

A/AC. 287/2016/PC. 2/L. 2　Provisional programme of work for the second session

A/AC. 287/2017/PC. 3/1　Agenda of the third session

A/AC. 287/2016/PC. 3/L. 1　Provisional agenda of the third session

A/AC. 287/2017/PC. 3/L. 2　Provisional programme of work for the third session

A/AC. 287/2017/PC. 4/1　Agenda of the fourth session

A/AC. 287/2016/PC. 4/L. 1　Provisional agenda of the fourth session

A/AC. 287/2017/PC. 4/L. 2　Provisional programme of work for the fourth session

主席对讨论的协助

一、导言

1. 根据大会第 72/249 号决议，正在召开的政府间会议将审议大会第 69/292 号决议所设预备委员会根据《联合国海洋法公约》的规定就国家管辖范围以外区域海洋生物多样性的养护和可持续利用问题拟订具有法律约束力的国际文书的内容提出的建议，并将拟订国际文书案文，以期尽早制定该文书（见第 72/249 号决议，第 1 段）。

2. 谈判将讨论 2011 年商定的一揽子内容所确定的专题，即国家管辖范围以外区域海洋生物多样性的养护和可持续利用，特别是作为一个整体的全部海洋遗传资源的养护和可持续利用，包括惠益分享问题，以及包括海洋保护区在内的划区管理工具、环境影响评估、能力建设和海洋技术转让等措施（同上，第 2 段）。

3. 政府间会议的工作和成果应完全符合《公约》的规定。该进程及其结果不应损害现行有关法律文书和框架以及相关的全球、区域和部门机构（同上，第 6 和 7 段）。

4. 2018 年 4 月 16 日至 18 日举行组织会议后，为讨论组织事项，包括拟订文书预稿的程序，政府间会议主席编写了本文件，以回应组织会议上提出的如下要求：借鉴预备委员会报告（A/AC. 287/2017/PC. 4/2）并注意到与该报告中第三 . A 节和第三 . B

节有关的建议，编写一份简明文件以协助讨论（同上，第 38 段）。会议同意，预备委员会编写的其他材料也将予以考虑。本文件的目的是引导会议转向编写文书预稿（见 A/CONF. 232/2018/2）。

5. 正如政府间会议所商定的，本文件不包含任何条约案文。本文件在报告第三 . A 节和第三 . B 节的基础上，确定了与一揽子内容全部要点和共有问题有关，需要进一步讨论的问题，包括可能需要处理的若干问题，包括在某些情况下可能涉及的备选方案（同上）。

6. 普遍的理解是，政府间会议第一届实质性会议的重点应是第 72/249 号决议规定的一揽子内容，并围绕一揽子内容的四个专题群组进行讨论（同上），有鉴于此，本文件将侧重于这些专题群组。本文件沿用第三 . A 节的结构，并且，除序言部分的内容、适用范围、财政资源和财务事项、遵守、争端解决、职责和责任、审查和最后条款之外，还在每个专题群组末尾增加了共有问题，以便确定共有问题与具体专题群组之间的实际关联。本文件的结构不影响未来文书的结构。

7. 在本文件中列入相关问题和备选方案并不意味着各代表团对这些问题和备选方案所涉各方面有着一致或趋同的意见。在提出了备选方案的情况下，这些备选方案的次序不应解释为表明关于优先次序的意向。

8. 请各代表团考虑对各种问题和备选方案的答复可能会产生的实际后果，特别是考虑如何在体现文书中。

9. 本文件的内容不影响任何代表团对其中所提任何事项的立场。此外，本文件所列要点、问题和备选方案未必详尽无遗，也不排除今后可能会审议未列入本文件的事项。

二、需进一步讨论的事项、问题和备选方案

10. 下文列出了政府间会议根据《公约》拟订一项关于养护和可持续利用国家管辖范围以外区域海洋生物多样性的具有法律约束力的国际文书案文时可以进一步审议的一些事项、问题和备选方案。

11. 这些事项、问题和备选方案依据的是预备委员会报告第三. A 节所列、多数代表团在预备委员会会议上意见趋同的非排他性要点，以及该报告第三 . B 节所列的一些存在意见分歧的主要事项。

12. 为便于参考，各章节和分节的编号均采用预备委员会报告第三 . A 节所用编号。因此，下文关于海洋遗传资源的第一部分，包括惠益分享问题，对应上述报告的第三 . A. 3 节；关于能力建设和海洋技术转让的最后一部分，对应该报告的第三 . A. 6 节。

13. 正如上文第 6 段所述，在本文件中，四个专题群组中每一个的末尾都增加了事项、问题和备选方案，对应上述报告第三 . A 节中的以下分节：第二分节，一般性要点（1. 用语；3. 目标；4. 与《公约》、其他文书和框架以及相关全球、区域和部门机构的关系）；第三分节，国家管辖范围以外区域海洋生物多样性的养护和可持续利用（1. 一般原则和方法；2. 国际合作）；第四分节，制度安排；第五分节，信息交换机制。还应指出，能力建设和海洋技术转让是应当纳入文书一揽子内容的各要点之中，还是列为一个专门部分并与其他部分相关联，或者采取其他办法，需要进一步讨论。

14. 如上文第 6 段所述，本文件没有处理预备委员会报告第三. A 节所列的以下分节：第一分节，序言要点；第二. 2 分节，适用范围；第六分节，财政资源和财务事项；第七分节，遵守；

第九分节，职责和责任；第十分节，审查；第十一分节，最后条款。这并不意味着这些要点将被排除在外；相反，随后将着手处理这些要点。

三、国家管辖范围以外区域海洋生物多样性的养护和可持续利用

3. 海洋遗传资源，包括惠益分享问题

虑及预备委员会报告第三节所列要点，可以考虑下文所列的一份不完全的事项、问题和备选方案清单：

3.1　范围

（a）文书采取什么方式规定这一章节的地域适用范围。范围是否涵盖：

（一）“区域”和公海的海洋遗传资源，或者“区域”或公海的海洋遗传资源？

（二）跨越国家管辖范围以内区域和（或）与上述区域重叠的海洋遗传资源？

（b）文书采取什么方式体现尊重沿海国对其国家管辖范围内的所有区域，包括对 200 海里以内和以外的大陆架和专属经济区的权利和管辖权。

（c）文书采取什么方式规定这一章节的属事适用范围。有待考虑的内容包括：

（一）文书是否应当区分使用鱼类和其他生物资源用于研究其遗传特性与作为商品这两种情况？作出区分的实际后果是什么？

（二）除了现场收集的海洋遗传资源之外，文书是否还将适用于不在原产地的海洋遗传资源，是否适用于由计算机模拟的海洋遗传资源和数字序列数据？这些选项的实际后果是什么？

（三）文书是否适用于衍生物？

3.2　获取和惠益分享

3.2.1　获取

（a）文书将采取什么方式处理获取问题，包括是否对国家管辖范围以外区域海洋遗传资源的获取进行规制。

（b）如果对获取进行规制：

（一）如何规制获取行为？

（二）这种规制的实际后果是什么，文书是否处理这些后果？如果是，如何处理？

（三）是否针对海洋遗传资源的取得地或来源地，规定不同的获取条款？

（四）是否对所有活动中的海洋遗传资源获取进行规制？

（c）如果获取不受规制，实际后果是什么，文书是否处理这些后果？如果是，如何处理？

3.2.2　惠益分享

（一）目标

在预备委员会报告第三节所列的惠益分享目标之外，还有什么目标（如有）可以列入文书？

（二）惠益分享的指导原则和方法

（a）在预备委员会报告第三节所列的惠益分享指导原则和方法之外，还有哪些指导原则和方法（如有）可以列入文书？需要进一步讨论人类共同继承财产和公海自由。

（b）文书应明确列出各项惠益分享指导原则和方法，还是在有关惠益分享的文书条款中落实这些指导原则和方法？

（三）惠益

（a）文书是否会载有一份惠益清单并且（或者）具体规定惠益类别？

（b）文书是否会规定以后拟订惠益清单？

（c）是否会对清单和（或）惠益类别进行审查？

(四) 惠益分享模式

(a) 惠益分享所需的实际安排，以及这些安排如何运作。有待考虑的问题包括：

(一) 文书是否将纳入一些条款，列出可能在不同阶段累积的惠益？

(二) 可以要求谁分享惠益？

(三) 谁可以成为受益人？

(四) 分享的惠益如何使用？

(b) 文书是否会依据取得或产生海洋遗传资源的地点，规定不同的惠益分享条款？

(c) 就惠益分享模式而言，需要考虑到哪些现有文书和框架？

(d) 如果规定惠益分享的信息交换机制，该机制需要包含哪些功能？

(e) 文书中还可以规定哪些其他惠益分享模式？

(f) 惠益分享模式应如何考虑到发展中国家的特殊情况，特别是最不发达国家、内陆发展中国家和地理条件不利的国家和小岛屿发展中国家以及非洲沿海国的特殊情况？

(g) 文书对惠益分享模式的规定应达到怎样的详细程度？

3.2.3　知识产权

文书是否将规定其与知识产权之间的关系？如果是，如何规定？

3.3　监测国家管辖范围以外区域海洋遗传资源的利用

(a) 文书将如何处理监测国家管辖范围以外区域海洋遗传资源的利用？

(b) 可以制定哪些切实安排（如有），以监测海洋遗传资源的利用，包括由谁负责实施这样的监测？

3.4　共有要点所涉问题

3.4.1　用语

海洋遗传资源，包括惠益分享问题有哪些关键用语的定义

（如有）将被纳入文书？

3.4.2　与《公约》、其他文书和框架以及相关全球、区域和部门机构的关系

对于一揽子内容的这一要点，是否需要一个专门条款来规定文书与《公约》、其他文书和框架以及相关全球、区域和部门机构的关系？

3.4.3　一般原则和方法

（a）除了在上文第 3.2.2（二）项之下可能考虑的内容之外，文书还可以纳入哪些与海洋遗传资源包括惠益分享问题有关的一般原则和方法？

（b）就海洋遗传资源，包括惠益分享问题而言，文书将采取什么最佳方式落实既定的一般原则和方法？

3.4.4　国际合作

文书应如何规定各国就海洋遗传资源，包括惠益分享问题开展合作的义务？

3.4.5　制度安排

（a）关于海洋遗传资源，包括惠益分享问题，是否需要具体的制度安排，同时考虑到利用现有机构、制度和机制的可能性？

（b）制度安排在海洋遗传资源，包括惠益分享问题上将发挥什么功能？

3.4.6　信息交换机制

（a）文书将规定哪些模式，以促进与海洋遗传资源，包括惠益分享问题有关的信息交换？

（b）除了预备委员会报告第三节所述的信息交换机制各项功能以及在上文第 3.2.2（四）项之下可能考虑的功能之外，与海洋遗传资源包括惠益分享问题有关的信息交换机制还有哪些其他功能（如有）可列入文书？关于海洋遗传资源包括惠益共享问题，还有什么其他资料可以传播？

（c）可以建立数据储存库等哪些其他机制？

（d）为了让数据储存库或信息交换机制等相关机制发挥所要求的功能，有哪些切实安排需要列入文书？

（e）可以考虑哪些现有的文书、机制和框架？

4. 划区管理工具包括海洋保护区等措施

可结合预备委员会报告第三节所含要点，考虑以下不完全清单所列事项、问题和备选方案：

4.1　划区管理工具包括海洋保护区的目标

如何在文书中纳入划区管理工具包括海洋保护区的特有目标？这些目标是否适合所有划区管理工具，包括海洋保护区？

4.2　与相关文书、框架和机构所规定措施的关系

（a）文书如何规定该文书项下措施与现行的相关法律文书和框架以及相关的全球、区域和部门机构所规定措施之间的关系。

（b）为解决文书所规定措施与毗邻沿海国所规定措施之间的兼容问题而纳入的条款。这些条款是否包含例如信息共享和（或）协商的规定？

（c）尊重沿海国对其国家管辖范围内所有区域，包括对200海里以内和以外的大陆架和专属经济区的权利，这点在文书中如何体现？

4.3　划区管理工具包括海洋保护区的有关程序

在划区管理工具包括海洋保护区方面，特别是决策和制度安排上，最宜采用何种程序，以增进合作与协调，同时避免损害现行法律文书和框架以及区域机构和部门机构的授权任务？

可能的办法包括全球办法、区域办法、部门办法和混合办法。

（a）就以上每种可能的办法和其他任何提议的办法而言，在划区管理工具包括海洋保护区方面、进而在区域确定、提案协商和评估、决策、执行、监测和审查方面，职能和责任拟议如何分配？

（b）为落实上文4.3.（a）提及的职能和责任分配拟议办法，

文书将作出哪些制度安排？

（c）为落实上文 4. 3.（a）提及的职能和责任分配拟议办法，文书将作出哪些切实安排？

（d）文书是否还将说明各种可能的办法如何适用于不同类型的划区管理工具？

4. 3. 1 确定区域

（a）考虑到上文 4. 3 提及的可能办法，文书将规定何种程序，用于依据现有的最佳科学资料、标准和准则，确定可能需要保护的区域？

（b）除了预备委员会报告第三节所含内容之外，文书还将纳入哪些标准和准则？如何兼顾相关国际、区域和部门机构使用的现行准则？

（c）文书在规定标准和准则时将达到怎样的详细程度？

（d）文书是否规定有可能审查和（或）更新标准和准则？

4. 3. 2 指定程序

（一）提案

（a）考虑到上文 4. 3 提及的可能办法以及预备委员会报告第三节所含的海洋保护区和其他有关的划区管理工具提案要点，文书还可纳入哪些其他要点？可以考虑的要点包括：

（一）谁可以提出提案？

（二）向谁递交提案？

（三）提案的内容，包括拟议措施的期限。

（二）就提案进行协商和评估

（a）考虑到上文 4. 3 提及的可能办法，文书是否具体规定将要参与协调和协商程序的利益攸关方？如是，纳入哪些利益攸关方？

（b）文书将纳入哪些提案协调和协商模式？

（c）文书将纳入哪些就提案出具科学咨询意见的模式？

（三）决策

（a）考虑到上文4.3提及的可能办法，在决策和制度安排上：

（一）文书将具体规定哪些决策模式，用于划区管理工具包括海洋保护区相关事项？

（二）为落实划区管理工具包括海洋保护区相关事项决策责任的拟议分配方案，文书将纳入哪些条款，包括制度安排？

（b）在何种基础上进行决策，才能增进合作与协调，同时避免损害现行法律文书和框架以及区域机构和部门机构的授权任务？

（c）文书应如何体现毗邻沿海国对决策程序的参与？

4.4　执行

考虑到上文4.3提及的可能办法，文书将纳入哪些条款，对该文书缔约方在适用于特定区域的措施方面所负责任作出规定？

4.5　监测和审查

考虑到上文4.3提及的可能办法，文书将如何规定评估划区管理工具包括海洋保护区的有效性以及之后采取后续行动，同时注意有必要采取适应性办法。

（a）评估由谁负责？

（b）后续行动由谁决定？

4.6　共有要点所涉问题

4.6.1　用语

关于划区管理工具，包括海洋保护区，有哪些关键用语的定义（如有）可以纳入文书？

4.6.2　与《公约》以及其他文书、框架和相关全球、区域和部门机构的关系

关于划区管理工具包括海洋保护区，除了在上文第4.3分节项下可能考虑的内容之外，还有哪些具体方面（如有）将会纳入文书？

4.6.3　一般原则和方法

（a）文书将纳入哪些关于在国家管辖范围以外区域使用划区

管理工具包括海洋保护区的一般原则和方法？

（b）文书怎样才能以最佳方式落实划区管理工具包括海洋保护区的一般原则和方法？

4.6.4 国际合作

文书将如何规定国家就划区管理工具包括海洋保护区开展合作的义务？

4.6.5 制度安排

（a）针对划区管理工具，包括海洋保护区，是否需要作出具体的制度安排，同时考虑到是否有可能利用现有的机构、制度和机制？

（b）对于划区管理工具，包括海洋保护区，制度安排将发挥哪些作用？

4.6.6 信息交换机制

（a）文书将规定哪些模式，用于促进划区管理工具包括海洋保护区方面的信息交流？

（b）除了预备委员会报告第三节所述的信息交换机制各项功能之外，信息交换机制在划区管理工具包括海洋保护区方面还有哪些其他功能（如有）可以纳入文书？关于划区管理工具，包括海洋保护区，还要传播哪些其他信息？

（c）还可以建立数据储存库等其他哪些机制？

（d）文书需要为数据储存库或信息交换机制等各种机制作出哪些切实安排，才能实现所需的功能？

（e）可以考虑哪些现有文书、机制和框架？

5. 环境影响评价

可结合预备委员会报告第三节所列要点，考虑下文不完全清单所列事项、问题和备选方案：

5.1 进行环境影响评价的义务

文书如何规定各国有义务就其管辖或控制下计划开展的活动

对国家管辖范围以外区域的潜在影响进行评估?

5.2　与相关文书、框架和机构的环境影响评价程序的关系

文书如何规定其与相关法律文书和框架以及相关全球、区域和部门机构的环境影响评价程序之间的关系。

5.3　需要进行环境影响评价的活动

(a) 文书将纳入环境影响评价的哪些阈值和准则? 具体如何体现?

(b) 是否制定一份清单，列明需要或不需要环境影响评价的活动，作为对阈值和准则的补充?

(c) 是否考虑累积影响? 如是，文书如何规定纳入考虑的累积影响?

(d) 文书是否纳入一个具体条款，要求对经认定在生态或生物方面具有重要意义或者脆弱性的区域实施环境影响评价?

5.4　环境影响评价程序

(a) 考虑到预备委员会报告第三节所述的环境影响评价程序的流程步骤，文书将纳入哪些流程步骤? 是否可以纳入任何其他步骤?

(b) 文书在环境影响评价的流程步骤上要达到怎样的详细程度?

(c) 关于环境影响评价程序，包括某项活动继续与否的决定，在何种程度上由国家完成或“国际化”? 如应“国际化”，程序的哪些方面应当“国际化”?

(d) 文书将如何体现毗邻沿海国的参与，例如何时参与、怎样参与?

5.5　环境影响评价报告的内容

(a) 考虑到预备委员会报告第三节关于环境影响评价报告所需内容的要点说明，文书将纳入环境影响评价报告的哪些内容? 是否可以纳入任何其他内容?

（b）文书在环境影响评价报告的内容上要达到怎样的详细程度？

（c）在处理跨境影响时，将采用以活动为导向的办法（立足于活动地点），以影响为导向的办法（立足于受影响的地点），还是两者相结合的办法？还可以考虑哪些其他办法（如有）？

5.6 监测、报告和审查

文书如何规定相关义务，以确保对授权开展的活动在国家管辖范围以外区域产生的影响进行监测、报告和审查？有待考虑的事项包括：

（a）监测、报告和审查程序在何种程度上由国家完成或“国际化”？如果该程序应“国际化”，则：

（一）监测、报告和审查的义务由谁承担？

（二）报告向谁提交？

（b）哪些信息要向毗邻沿海国家提供？怎样以及何时提供信息？

5.7 战略环境评价

文书是否纳入关于战略环境评价的条款？如是：

（a）评价范围是什么？

（b）关于国家管辖范围以外区域海洋生物多样性的战略环境评价，将在全球一级还是区域一级实施？

（c）谁负责实施战略环境评价？

（d）如何根据战略环境评价的结果采取后续行动？

5.8 共有要点所涉问题

5.8.1 用语

环境影响评价有哪些关键用语定义（如有）可以纳入文书？

5.8.2 与《公约》以及其他文书、框架和相关全球、区域和部门机构的关系

关于环境影响评价，除了在上文第5.2分节项下可能考虑的

内容之外，还有哪些具体方面将会纳入文书?

5.8.3　一般原则和方法

（a）关于环境影响评价，文书将纳入哪些一般原则和方法?

（b）文书怎样才能以最佳方式落实环境影响评价的一般原则和方法?

5.8.4　国际合作

文书如何规定各国就环境影响评价开展合作的义务?

5.8.5　制度安排

（a）针对环境影响评价，是否需要作出具体的制度安排，同时考虑到是否有可能利用现有的机构、制度和机制?

（b）对于环境影响评价，制度安排将发挥哪些作用?

5.8.6　信息交换机制

（a）文书将规定哪些模式，以促进环境影响评价方面的信息交流?

（b）除了预备委员会报告第三节所述的信息交换机制各项功能之外，关于环境影响评价相关信息交换机制，还有哪些其他功能可以纳入文书?环境影响评价还有哪些其他信息需要传播?

（c）可以建立数据储存库等哪些其他机制?

（d）文书需要为数据储存库或信息交换机制等各种机制作出哪些切实安排，才能实现所要求的功能?

（e）可以考虑哪些现有文书、机制和框架?

6. 能力建设和海洋技术转让

可结合预备委员会报告第三节所含要点，考虑到以下不完全清单所列事项、问题和备选方案：

6.1　能力建设和海洋技术转让的目标

（a）采取什么方式在文书中纳入能力建设和海洋技术转让的目标?

（b）承认发展中国家，特别是最不发达国家、内陆发展中国家、地理不利国和小岛屿发展中国家以及非洲沿海国家的特殊要求，在文书中如何体现？

（c）文书如何依照《公约》第 266 条第 2 款的规定，处理并体现发展和加强各国能力的需求，尤其是对有这种需要和要求的发展中国家？

6.2 能力建设和海洋技术转让的类别和模式

（a）在现有文书，例如《公约》和政府间海洋学委员会的《海洋技术转让标准和准则》的基础上，文书是否纳入一份指示性不完全清单，列出能力建设和海洋技术转让的大类类型？

（一）如纳入上述清单：

- 清单如何编制？由谁编制？如何更新？
- 编制清单时，还可借鉴哪些其他文书？
- 清单的范围有多大？

（二）如果文书不含清单：

- 文书是否规定之后将编制一份清单？
- 能力建设和海洋技术转让的类别还可通过何种方式得到体现？

（b）文书将纳入哪些与海洋遗传资源包括惠益分享问题、划区管理工具包括海洋保护区等措施以及环境影响评价有关的具体合作和援助形式？

（c）文书将纳入哪些能力建设和海洋技术转让模式？

（一）鉴于预备委员会报告第三节列举了能力建设和海洋技术转让模式的可能参数，问题是文书将规定能力建设和海洋技术转让模式的哪些参数。除其他外，参数还涉及能力建设和技术转让的提供者和提供基础。

（二）是利用现行机制，还是建立新机制？

（d）文书将为海洋技术转让规定哪些条款和条件？该等条款和条件如何兼顾现有文书？

（e）除了预备委员会报告第三节所述的信息交换机制可能具

有的各项功能之外，信息交换机制在能力建设和海洋技术转让方面还有哪些其他功能（如有）可以纳入文书？还有哪些与能力建设和海洋技术转让有关的其他信息将会通过信息交换机制传播？谁能使用这种信息交换机制？

(f) 就信息交换机制的功能而言，文书将考虑哪些组织的工作？

(g) 如何审查能力建设和海洋技术转让模式？

6.3　供资

(a) 在资金和资源提供方面，需要考虑哪些现有机制？

(b) 文书如何在考虑现有机制的情况下，对资金和资源的提供作出规定？是否考虑：

(一) 谁能获得资金和资源？

(二) 谁来提供资金和资源？

(三) 如何使用资金和资源？

(四) 文书将如何处理资金和资源的持续性、可预测性和可获取性问题？

6.4　监测和审查

文书如何处理监测和审查能力建设和海洋技术转让活动有效性的问题，如何规定可能采取的后续行动？有待考虑的事项包括：

(a) 谁负责监测和审查？

(b) 监测和审查的内容是什么？

(c) 如何根据监测和审查情况采取后续行动？

6.5　共有要点所涉问题

6.5.1　用语

能力建设和海洋技术转让有哪些关键用语的定义（如有）可以纳入文书？

6.5.2　与《公约》以及其他文书、框架和相关全球、区域和部门机构的关系

对于一揽子内容中的这个要点，是否需要一个具体条款规定

文书与《公约》、其他文书和框架以及相关全球、区域和部门机构的关系？

6. 5. 3　一般原则和方法

（a）文书将纳入哪些与能力建设和海洋技术转让有关的一般原则和方法？

（b）文书怎样才能以最佳方式落实能力建设和海洋技术转让的一般原则和方法？

6. 5. 4　国际合作

文书将如何规定国家在能力建设和海洋技术转让方面开展合作的义务？

6. 5. 5　制度安排

（a）能力建设和海洋技术转让是否需要具体的制度安排，同时考虑到有可能利用现有的机构、制度和机制？

（b）制度安排将在能力建设和海洋技术转让方面发挥哪些功能？

6. 5. 6　信息交换机制

（a）文书将设置哪些模式，以促进能力建设和海洋知识转让方面的信息交流？

（b）除了上文第 6. 2 分节提到的信息交换机制之外，还可建立数据储存库等其他哪些机制？

（c）文书需要为数据储存库或信息交换机制等各种机制作出哪些切实安排，才能实现所要求的功能？

（d）可以考虑哪些现有文书、机制和框架？

President's aid to discussions

Ⅰ. Introduction

1. Theintergovernmental conference is being convened pursuant to General Assembly resolution 72/249 to consider the recommendations of the Preparatory Committee established pursuant to Assembly resolution 69/292 on the elements and to elaborate the text of an international legally binding instrument under the United Nations Convention on the Law of the Sea on the conservation and sustainable use of marine biological diversity of areas beyond national jurisdiction, with a view to developing the instrument as soon as possible (see resolution 72/249, para. 1).

2. Thenegotiations shall address the topics identified in the package agreed in 2011, namely, the conservation and sustainable use of marine biological diversity of areas beyond national jurisdiction, in particular, together and as a whole, marine genetic resources, including questions on the sharing of benefits, measures such as area-based management tools, including marine protected areas, environmental impact assessments and capacity-building and the transfer of marine technology (ibid., para. 2).

3. The work andresults of the conference should be fully consistent with the provisions of the Convention. The process and its result should

not undermine existing relevant legal instruments and frameworks and relevant global, regional and sectoral bodies (ibid., paras. 6-7).

4. Followingthe organizational meeting, held from 16 to 18 April 2018, to discuss organizational matters, including the process for the preparation of the zero draft of the instrument, the present document was prepared by the President of the conference in response to the request by the conference at that meeting, to prepare a concise document as an aid to discussions, building on the report of the Preparatory Committee (A/AC. 287/2017/PC. 4/2) and bearing in mind the recommendations concerning sections III. A and III. B of the report (ibid., para. 38). As agreed by the conference, other materials produced in the context of the Preparatory Committee were also considered. The present document is aimed at putting the conference on a path to the preparation of a zero draft of the instrument (seeA/CONF. 232/2018/2).

5. Asagreed by the conference, the document does not contain any treaty text. Rather, it identifies, on the basis of sections III. A and III. B of the report, issues that need to be further discussed in respect of all elements of the package and cross-cutting issues, and includes a limited number of possible questions to be addressed, including, in some cases, possible options in relation thereto (ibid.).

6. In thelight of the general understanding that the first substantive session of the conference should focus on the elements of the package as set out in resolution 72/249 and that the discussions be organized around the four thematic clusters of the package (ibid.), the present document focuses on those thematic clusters. The structur e of section III. A has been maintained and cross-cutting issues, apart from the preambular elements, scope of application, financial resources and issues, compliance, the settlement of disputes, responsibility and liability, review

and final clauses, have been added to the end of each thematic cluster, with a view to facilitating the determination of how those cross-cutting issues might relate, in practical terms, to those specific clusters. The structure of the present document is without prejudice to the structure of the future instrument.

7. Theinclusion of questions and options herein does not imply that there was agreement or a convergence of views among delegations concerning the aspects to which those questions and options relate. Where options are presented, the order of such options should not be construed as indicating a suggested order of priority.

8. Delegationsare invited to consider the practical consequences of the answers to various questions and options and, in particular, how they could be reflected in the instrument.

9. Thecontent of the present document is without prejudice to the position of any delegation on any of the matters referred to herein. Furthermore, the elements, questions and options listed are not necessarily exhaustive and do not preclude the consideration of matters that have not been included in the document.

Ⅱ. Issues, questions and options to be further discussed

10. Some ofthe issues, questions and options that may be further considered by the conference in the elaboration of the text of an international legally binding instrument under the Convention on the conservation and sustainable use of marine biological diversity of areas beyond national jurisdiction, with a view to developing the instrument as soon as possible, are set out below.

11. Theissues, questions and options are based on the non-exclusive elements that generated convergence among most delegations at the meeting of the Preparatory Committee, as reflected in section III. A of its report, and some of the main issues on which there was a divergence of views, as reflected in section III. B of the report.

12. Forease of reference, the numbering of the sections and subsections is based on that used in section III. A of the report of the Preparatory Committee. Therefore, the first section presented below, on marine genetic resources, including questions on the sharing of benefits, corresponds to section III. A. 3 of the report, and the last section, on capacity-building and the transfer of marine technology, corresponds to section III. A. 6 of the report.

13. Aspreviously noted in paragraph 6 above, issues, questions and options have been added to the end of each of the four thematic clusters in the present document, corresponding to the following subsections in section III. A of the report: subsection II, General elements (1. Use of terms; 3. Objectives; and 4. Relationship to the Convention and other instruments and frameworks and relevant global, regional and sectoral bodies); subsection III, Conservation and sustainable use of marine biological diversity of areas beyond national jurisdiction (1. General principles and approaches; and 2. International cooperation); subsection IV, Institutional arrangements; and subsection V, Clearing-house mechanism. It is also noted that the question of whether to mainstream capacity-building and the transfer of marine technology across the various elements of the package in the instrument, to include it in a dedicated section with links to the other sections, or to adopt a different approach, requires further discussion.

14. Asalso indicated in paragraph 6 above, the following subsections

under section III. A of the report of Preparatory Committee have not been addressed in the present document: subsection I, Preambular elements; subsection II. 2, Scope of application; subsection VI, Financial resources and issues; subsection VII, Compliance; subsection VIII, Settlement of disputes; subsection IX, Responsibility and liability; subsection X, Review; and subsection XI, Final clauses. This does not mean that these elements will be excluded; rather, they will be taken up subsequently.

Ⅲ. Conservation and sustainable use of marine biological diversity ofareas beyond national jurisdiction

3. Marine geneticresources, including questions on the sharing of benefits

Bearing in mind the elements reflected in section III of the report of the Preparatory Committee, a non-exhaustive list of issues, questions and options may be considered, as follows:

3. 1 Scope

(a) The manner in which the geographical scope of application of this section would be reflected in the instrument. Would the scope cover marine genetic resources:

(ⅰ) Of the Area and the high seas, or of the Area or the high seas?

(ⅱ) That straddleand/or overlap with areas within national jurisdictions?

(b) The manner in which respect for the rights and jurisdiction of coastal States over all areas under their national jurisdiction, including the continental shelf within and beyond 200 nautical miles and the exclusive economic zone, would be reflected in the instrument.

(c) The manner in which the material scope of application of this section of the instrument would be reflected in the instrument. Elements to consider may include:

(ⅰ) Would a distinction be made in the instrument between use of fish and other biological resources for research into their genetic properties and their use as a commodity? What would be the practical consequences of such a distinction?

(ⅱ) Otherthan marine genetic resources collected *in situ*, would the instrument also apply to *ex situ* marine genetic resources and to *in silico* marine genetic resources and digital sequence data? What would be the practical consequences of these options?

(ⅲ) Would the instrument apply to derivatives?

3.2 Accessand benefit-sharing

3.2.1 Access

(a) The mannerin which access would be addressed in the instrument, including whether access to the marine genetic resources of areas beyond nationa l jurisdiction would be regulated.

(b) Ifaccess is regulated:

(ⅰ) How would accessbe regulated?

(ⅱ) What would be the practical consequences of such regulation and would they be addressed in the instrument? If so, how?

(ⅲ) Would there be different access provisions depending on where the marine genetic resources are sourced or originate from?

(ⅳ) Would access to marine genetic resources be regulated for all activities?

(c) Ifaccess is unregulated, what would be the practical consequences and would they be addressed in the instrument? If so, how?

3.2.2 Sharingof benefits

(i) *Objectives*

What objectives of the sharing of benefits, if any, in addition to those included in section III of the report of the Preparatory Committee, could be included in the instrument?

(ii) *Principlesand approaches guiding benefit-sharing*

(a) Which principlesand approaches guiding benefit-sharing, if any, in addition to those included in section III of the report of the Preparatory Committee, could be included in the instrument? Further discussions are required with regard to the common heritage of mankind and the freedom of the high seas.

(b) Would the principles and approaches guiding benefit-sharing be explicitly listed in the instrument or would they be operationalized in the provisions of the instrument concerning benefit-sharing?

(iii) *Benefits*

(a) Would the instrument contain a list of benefits and/or would it specify types of benefits?

(b) Would the instrument provide for a list to be developed subsequently?

(c) Would there be a review of any such list and/or of the types of benefits?

(iv) *Benefit-sharing modalities*

(a) The practical arrangements that would be required for the sharing of benefits and how these would be operationalized. Issues to consider may include:

(i) Would the instrument include provisions setting out benefits that might accrue at different stages?

(ⅱ) Who might berequired to share benefits?

(ⅲ) Who might bethe beneficiaries?

(ⅳ) How might the shared benefits be used?

(b) Would the instrument include different provisions on the sharing of benefits depending on where the marine genetic resources are sourced or originate from?

(c) What existing instruments and frameworks would need to be taken into account with regard to modalities for the sharing of benefits?

(d) If a clearing-house mechanism is provided for the sharing of benefits, what functions would it need to include?

(e) What other modalitiesfor the sharing of benefits could be provided for in the instrument?

(f) How would the special circumstances of developing countries, in particular the least developed countries, landlocked developing countries, geographically disadvantaged States and small island developing States, as well as coastal African States, be taken into account in the modalities for the sharing of benefits?

(g) How much detail on the modalities for the sharing of benefits would be included in the instrument?

3.2.3 Intellectual property rights

Would the relationship between the instrument and intellectual property rights be set out in the instrument? If so, how?

3.3 Monitoringof the utilization of marine genetic resources of areas beyond national jurisdiction

(a) How couldthe instrument address the monitoring of the utilization of marine genetic resources of areas beyond national jurisdiction?

(b) What practical arrangements, if any, could be developed to monitor the utilization of marine genetic resources, including who would be responsible for undertaking such monitoring?

3.4 Issues fromthe cross-cutting elements

3.4.1 Use of terms

Which definitions of key terms pertaining to marine genetic resources, including questions on the sharing of benefits, if any, would be included in the instrument?

3.4.2 Relationshipto the Convention and other instruments and frameworks and relevant global, regional and sectoral bodies

Would this element of the package require a specific provision on the relationship to the Convention, other instruments and frameworks, and relevant global, regional and sectoral bodies?

3.4.3 General principlesand approaches

(a) Which general principlesand approaches pertaining to marine genetic resources, including questions on the sharing of benefits, in addition to what might be considered in the context of subsection 3.2.2 (ii) above, could be included in the instrument?

(b) How would the instrument best give effect to the identified general principles and approaches in the context of marine genetic resources, including questions on the sharing of benefits?

3.4.4 International cooperation

How would the instrument set out the obligation of States to cooperate with respect to marine genetic resources, including questions on the sharing of benefits?

3.4.5 Institutional arrangements

(a) Inrespect of marine genetic resources, including questions on the sharing of benefits, would specific institutional arrangements be required, taking into account the possibility of using existing bodies, institutions and mechanisms?

(b) What functionswould institutional arrangements have in respect of marine genetic resources, including questions on the sharing of benefits?

3.4.6 Clearing-house mechanism

(a) What modalitieswould the instrument set out to facilitate the exchange of information relevant to marine genetic resources, including questions on the sharing of benefits?

(b) Inaddition to the functions for a clearing-house mechanism addressed in section III of the report of the Preparatory Committee, and what might be considered in the context of subsection 3.2.2 (iv) above, what other functions for a clearing-house mechanism in respect of marine genetic resources, including questions on the shari ng of benefits, if any, would be included in the instrument? What other information regarding marine genetic resources, including questions on the sharing of benefits, would be disseminated?

(c) Whatother mechanisms, such as data repositories, might be established?

(d) What practical arrangementswould need to be included in the instrument for mechanisms such as data repositories or a clearing-house mechanism in order to fulfil the required functions?

(e) Whatexisting instruments, mechanisms and frameworks could be taken into account?

4. Measures suchas area-based management tools, including marine protected areas

Bearing in mind the elements reflected in section III of the report of the Preparatory Committee, a non-exhaustive list of issues, questions and options may be considered, as follows:

4.1 Objectivesof area-based management tools, including marine protected areas

The manner in which objectives specific to area-based management tools, including marine protected areas, would be included in the instrument. Would these objectives apply to the full range of area-based management tools, including marine protected areas?

4.2 Relationshipto measures under relevant instruments, frameworks and bodies

(a) The manner in which the instrument would set out the relationship between measures under the instrument and measures under existing relevant legal instruments and frameworks and relevant global, regional and sectoral bodies.

(b) Theprovisions that would be included to address issues of compatibility between measures under the instrument and those established by adjacent coastal States. Would the provisions include, for example, provisions for the sharing of information and/or for consultation?

(c) The mannerin which the instrument would reflect respect for the rights of coastal States over all areas under their national jurisdiction, including the continental shelf within and beyond 200 nautical miles and the exclusive economic zone.

4.3 Processin relation to area-based management tools, including marine protected areas

What would be the most appropriate process in relation to area-based management tools, including marine protected areas, in particular with respect to decision-making and institutional set up, with a view to enhancing cooperation and coordination, while avoiding undermining existing legal instruments and frameworks and the mandates of regional and sectoral bodies?

Possible approaches might include a global approach, a regional approach, a sectoral approach and a hybrid approach.

(a) For each of these possible approaches and for any other proposed approach, what would be the proposed allocation of roles and responsibilities in relationto area-based management tools, including marine protected areas, including with respect to the identification of areas, consultation on and assessment of proposals, decision-making, implementation, and monitoring and review?

(b) What institutional arrangementswould be included in the instrument to give effect to the proposed allocation of roles and responsibilities under 4. 3. a above?

(c) What practical arrangementswould be included in the instrument to give effect to the proposed allocation of roles and responsibilities under 4. 3. a above?

(d) Would the instrument also address how the possible approaches would apply to the different types of area-based management tools?

4. 3. 1 Identificationof areas

(a) Taking into account possible approaches as indicated in 4. 3 above, what process for the identification of areas within which protection may be required, based on the best available scientific information, standards and criteria, would the instrument set out?

(b) Which standardsand criteria, in addition to those included in section III of the report of the Preparatory Committee, would be included in the instrument? How would existing criteria that are utilized by relevant global, regional and sectoral bodies be taken into account?

(c) How much detail would the instrument include in setting out the standards and criteria?

(d) Would the instrument provide for the possibility of reviewing and/or updating the standards and criteria?

4.3.2 Designation process

(ⅰ) *Proposal*

(a) Takinginto account possible approaches indicated in 4.3 above, as well as the elements of proposals related to marine protected areas, and other area - based management tools where relevant, included in section III of the report of the Preparatory Committee, what other elements would be included in the instrument? Elements to consider may include:

(ⅰ) Who can make proposals?

(ⅱ) Who wouldthe proposals be submitted to?

(ⅲ) Thecontent of the proposals, including the duration of the proposed measure.

(ⅱ) *Consultationon and assessment of the proposal*

(a) Takinginto account possible approaches as indicated in 4.3 above, would the instrument specify the stakeholders who would be involved in the coordination and consultations process? If so, which stakeholders would be included?

(b) What modalitiesfor coordination and consultations on the proposal would be included in the instrument?

(c) What modalitiesfor the provision of scientific advice on the proposal would be included in the instrument?

(ⅲ) *Decision-making*

(a) Takinginto account possible approaches as indicated in 4.3 above, with respect to decision-making and institutional set up:

(ⅰ) What modalitiesfor decision-making on matters related to area-based management tools, including marine protected areas, would be specified in the instrument?

(ii) What provisions, including any institutional arrangements, would the instrument include to give effect to the proposed allocation of responsibility for decision-making on matters related to area-based management tools, including marine protected areas?

(b) On what basis would decisions be made, with a view to enhancing cooperation and coordination, while avoiding undermining existing legal instruments and frameworks and the mandates of regional and sectoral bodies?

(c) How would the instrument reflect the involvement of adjacent coastal States in the decision-making process?

4.4 Implementation

Taking into account possible approaches as indicated in 4.3 above, what provisions would the instrument include to provide for the responsibility of parties to the instrument in relation to the measures for a particular area?

4.5 Monitoringand review

Takinginto account possible approaches as indicated in 4.3 above, the manner in which the instrument would provide for the assessment of the effectiveness of area-based management tools, including marine protected areas, and the subsequent follow-up actions that would be set out in the instrument, bearing in mind the need for an adaptive approach.

(a) Who would beresponsible for such assessments?

(b) Who coulddecide on the follow-up actions?

4.6 Issues fromthe cross-cutting elements

4.6.1 Use of terms

Which definitions of key terms pertaining to area-based

management tools, including marine protected areas, if any, would be included in the instrument?

4. 6. 2　Relationshipto the Convention and other instruments and frameworks and relevant global, regional and sectoral bodies

What specific aspects pertaining to area-based management tools, including marine protected areas, if any, in addition to what might be considered in the context of subsection 4. 3 above, would be included in the instrument?

4. 6. 3　General principlesand approaches

(a) Which general principlesand approaches pertaining to area-based management tools, including marine protected areas, in areas beyond national jurisdiction, could be included in the instrument?

(b) Howwould the instrument best give effect to the identified general principles and approaches in the context of area-based management tools, including marine protected areas?

4. 6. 4　International cooperation

How would the instrument set out the obligation of States to cooperate with respect to area-based management tools, including marine protected areas?

4. 6. 5　Institutional arrangements

(a) Wouldarea-based management tools, including marine protected areas, require specific institutional arrangements, taking into account the possibilit y of using existing bodies, institutions and mechanisms?

(b) What functionswould institutional arrangements carry out in respect of area-based management tools, including marine protected areas?

4. 6. 6　Clearing-house mechanism

(a) What modalitieswould the instrument set out to facilitate the exchange of information relevant to area-based management tools, including marine protected areas?

(b) Inaddition to the functions for a clearing-house mechanism in section III of the report of the Preparatory Committee, what other functions for a clearing-house mechanism in respect of area-based management tools, including marine protected areas, if any, would be included in the instrument? What other information regarding area-based management tools, including marine protected areas, would be disseminated?

(c) Whatother mechanisms, such as data repositories, might be established?

(d) What practical arrangementswould need to be included in the instrument for mechanisms such as data repositories or a clearing-house mechanism in order to fulfil the required functions?

(e) Whatexisting instruments, mechanisms and frameworks could be taken into account?

5. Environmental impact assessments

Bearing in mind the elements reflected in section III of the report of the Preparatory Committee, a non-exhaustive list of issues, questions and options may be considered, as follows:

5.1 Obligationto conduct environmental impact assessments

The manner in which the instrument would set out the obligation for States to assess the potential effects of planned activities under their jurisdiction or control in areas beyond national jurisdiction.

5.2 Relationshipto environmental impact assessment processes under relevant instruments, frameworks and bodies

The manner in which the instrument would set out its relationship to environmental impact assessment processes under other relevant legal instruments and frameworks and relevant global, regional and sectoral bodies.

5.3 Activities for which an environmental impact assessment is required

(a) The thresholds and criteria for environmental impact assessments that would be included in the instrument and how these would be reflected.

(b) Woulda list of activities that require or do not require an environmental impact assessment complement those thresholds and criteria?

(c) Would cumulative impacts be taken into account? If so, how would the instrument provide for such impacts being taken into account?

(d) Would the instrument include a specific provision for environmental impact assessments in areas identified as ecologically or biologically significant or vulnerable?

5.4 Environmentalimpact assessment process

(a) Taking into account the procedural steps of the environmental impact assessment process set out in section III of the report of the Preparatory Committee, which procedural steps would be included in the instrument? Are there any other steps that could be included?

(b) How much detail regarding procedural stepsfor environmental impact assessment would be included in the instrument?

(c) To what degree would the environmental impact assessment process, including the decision on whether an activity would proceed or not, be conducted by States or be "internationalized"? If the process were to be "internationalized", which aspects of the process should be "internationalized"?

(d) How would the instrument reflect the involvement of adjacent coastal States, for example, and when and how would such involvement take place?

5.5 Contentof environmental impact assessment reports

(a) Taking into account the elements in section III of the report of the Preparatory Committee with respect to the required content of environmental impact assessment reports, what components of environmental impact assessment reports would be included in the instrument? Are there any additional components that could be included?

(b) How much detail on the content of environmental impact assessment reports would be set out in the instrument?

(c) In addressing transboundary impacts, would an activity-oriented approach (based on the location of the activity), an impact-oriented approach (based on the location of the impact) or a combination of both be adopted? What other approaches, if any, could be considered?

5.6 Monitoring, reportingand review

The manner in which the instrument would set out the obligation to ensure that the impacts of authorized activities in areas beyond national jurisdiction are monitored, reported and reviewed. Issues to consider may include:

(a) To what degree would the monitoring, reporting and review process be conducted by States or be "internationalized"? If the process were to be "internationalized":

(i) Who would havethe obligation to monitor, report and review?

(ii) To whom would reports be submitted?

(iii) What informationwould be provided to adjacent coastal States and how and when would that information be communicated?

5.7 Strategic environmental assessments

Would the instrument include provisions on strategic environmental assessments? If so:

(a) What would be the scope of such assessments?

(b) Would strategic environmental assessments with respect to marine biological diversity of areas beyond national jurisdiction be conducted at the global or regional level?

(c) Who would beresponsible for conducting of strategic environmental assessments?

(d) How would the results of the strategic environmental assessments be followed-up on?

5.8 Issues fromthe cross-cutting elements

5.8.1 Use of terms

Which definitions of key terms pertaining to environmental impact assessments, if any, would be included in the instrument?

5.8.2 Relationshipto the Convention and other instruments and frameworks and relevant global, regional and sectoral bodies

What specificaspects pertaining to environmental impact assessments, if any, in addition to what might be considered in the context of subsection 5.2 above, would be included in the instrument?

5.8.3 General principlesand approaches

(a) Which general principlesand approaches pertaining to environmental impact assessments could be included in the instrument?

(b) How would the instrument best give effect to the identified general principles and approaches in the context of environmental impact assessments?

5.8.4 International cooperation

How would the instrument set out the obligation of States to cooperate with respect to environmental impact assessments?

5.8.5 Institutional arrangements

(a) Would environmental impact assessments require specific institutional arrangements, taking into account the possibility of using existing bodies, institutions and mechanisms?

(b) What functions would institutional arrangements carry out in respect of environmental impact assessments?

5.8.6 Clearing-house mechanism

(a) What modalitieswould the instrument set out to facilitate the exchange of information relevant to environmental impact assessments?

(b) Inaddition to the functions for a clearing-house mechanism in section III of the report of the Preparatory Committee, what other functions for a clearing-house mechanism in respect of environmental impact assessments, if any, would be included in the instrument? What other information regarding environmental impact assessments would be disseminated?

(c) What other mechanisms, such as data repositories, might be established?

(d) What practical arrangementswould need to be included in the instrument for mechanisms such as data repositories or a clearing-house mechanism in order to fulfil the required functions?

(e) Whatexisting instruments, mechanisms and frameworks could be taken into account?

6. Capacity-buildingand the transfer of marine technology

Bearing in mind the elements reflected in section III of the report of the Preparatory Committee, a non-exhaustive list of issues, questions and options may be considered, as follows:

6.1 Objectivesof capacity-building and the transfer of marine technology

(a) The manner in which the objectives of capacity-building and the transfer of marine technology would be included in the instrument.

(b) How would the instrument reflect the recognition of the special requirements of developing countries, in particular the least developed countries, landlocked developing countries, geographically disadvantaged States and small island developing States, as well as coastal African States?

(c) How would the instrument address and reflect the need to develop and strengthen the capacity of States, in particular developing States, that need and request it, in accordance with article 266 (2) of the Convention?

6.2 Types of and modalities for capacity-building and transfer of marine technology

(a) Drawingon existing instruments, such as the Convention and the Criteria and Guidelines on Transfer of Marine Technology of the Intergovernmental Oceanographic Commission, would the instrument include an indicative, non-exhaustive list of broad categories of types of capacity-building and transfer of marine technology?

(ⅰ) If a list were to be included:

• How would the list be developed and by whom? How would it be updated?

• What other instruments would be drawn from to develop such a list?

• How broadwould the list be?

(ⅱ) If no list were to be included in the instrument:

• Would the instrument provide for a list to be developed subsequently?

• Howelse could the types of capacity-building and transfer of marine technology be reflected?

(b) Whatspecific forms of cooperation and assistance would be included in the instrument in relation to marine genetic resources, including questions on the sharing of benefits, measures such as area-based management tools, including marine protected areas, and environmental impact assessments?

(c) Modalities for capacity-building and the transfer of marine technology that would be included in the instrument.

(i) Bearingin mind the possible parameters of modalities for capacity- building and the transfer of marine technology in section III of the report of the Preparatory Committee, the issue is what parameters the instrument would set out for the modalities regarding capacity-building and the transfer of marine technology. Parameters could also relate to, inter alia, who the providers of capacity-building and technology transfer would be and the basis on which capacity-building and technology transfer would be provided.

(ii) Would existing mechanisms be utilized or would new mechanisms be developed?

(d) What termsand conditions could the instrument set out for the transfer of marine technology? How would any such terms and conditions take into account existing instruments?

(e) In addition to the information set out in section III of the report of the Preparatory Committee with respect to possible functions of a clearing-house mechanism, what other functions for a clearing-house mechanism in respect of capacity-building and the transfer of marine technology, if any, would be included in the instrument? What other information or data, if any, relating to capacity-building and the transfer of marine technology, would be disseminated by a clearing-house mechanism? Who would have access to such a clearing-house mechanism?

(f) Which organizations' work would the instrument take into account with respect to the functions of a clearing-house mechanism?

(g) How would the modalities for capacity-building and transfer of marine technology be reviewed?

6.3 Funding

(a) The existing mechanisms that would need to be taken into account in the provision of funding and resources.

(b) The manner in which the instrument would address the provision of funding and resources, taking into account existing mechanisms. Would the instrument consider:

(ⅰ) Who would haveaccess to the funding and resources?

(ⅱ) Who would contribute funding and resources?

(ⅲ) How the funds and resources would be used?

(ⅳ) How the instrument would address the sustainability, predictability and accessibility of such funding and resources?

6.4 Monitoringand review

The manner in which the instrument would address the issue of monitoring and review of the effectiveness of capacity-building and the transfer of marine technology activities and possible follow-up action. Issues to consider may include:

(a) Who would undertake such monitoring and review?

(b) What would be the subject matter of any such monitoring and review?

(c) How would such monitoring and review be followed up on?

6.5 Issues fromthe cross-cutting elements

6.5.1 Use of terms

Which definitions of key terms pertaining to capacity-building and the transfer of marine technology, if any, could be included in the instrument?

6.5.2 Relationshipto the Convention and other instruments and frameworks and relevant global, regional and sectoral bodies

Would this element of the package require a specific provision on the relationship to the Convention, other instruments and frameworks, and relevant global, regional and sectoral bodies?

6.5.3 General principlesand approaches

(a) Which general principlesand approaches pertaining to capacity-building and the transfer of marine technology could be included in the instrument?

(b) How would the instrument best give effect to the identified general principles and approaches in the context of capacity-building and the transfer of marine technology?

6.5.4 International cooperation

How would the instrument set out the obligation of States to cooperate with respect to capacity-building and the transfer of marine technology?

6.5.5 Institutional arrangements

(a) Would capacity-building and the transfer of marine technology require specific institutional arrangements, taking into account the possibility of using existing bodies, institutions and mechanisms?

(b) What functionswould institutional arrangements have in respect of capacity-building and the transfer of marine technology?

6.5.6 Clearing-house mechanism

(a) What modalitiesto facilitate the exchange of information relevant to capacity-building and the transfer of marine technology would be included in the instrument?

(b) Other than the clearing-house mechanism referred to in subsection 6.2 above, what other mechanisms, such as data repositories, might be established?

(c) What practical arrangements would need to be included in the instrument for mechanisms such as data repositories or a clearing-house mechanism in order to fulfil the required functions?

(d) What existing instruments, mechanisms and frameworks could be taken into account?